Process Industry Manufacturing Software:

ERP, Planning, Recipe, MES & Process Control

Shaun Snapp

Process Industry Manufacturing Software: ERP, Planning, Recipe, MES & Process Control

For information about this title or to order other books and/or electronic media, contact the publisher:
SCM Focus Press
PO Box 29502 #9059
Las Vegas, NV 89126-9502
http://www.scmfocus.com/scmfocuspress
(408) 657-0249

ISBN: 978-1-939731-18-0

Printed in the United States of America

Cover and interior design by: 1106 Design

Contents

CHAPTER 1: Introduction . . . 1
CHAPTER 2: Continuous versus Batch Processing. . . . 21
CHAPTER 3: The Discrete Manufacturing Focus of Most Manufacturing Software . . . 85
CHAPTER 4: Co-Products and By-Products . . . 101
CHAPTER 5: Recipe Management. . . . 105
CHAPTER 6: Overlapping, Leading and Lagging Operations . . . 125
CHAPTER 7: Mixing/Blending/Pooling Operations . . . 131
CHAPTER 8: Changeovers and Cleaning . . . 137
CHAPTER 9: Constraint Planning and Process Control . . . 157
CHAPTER 10: Manufacturing Execution Scheduling. . . . 201
CHAPTER 11: ERP for Process Industry Manufacturing . . . 209
CHAPTER 12: Conclusion . . . 225
APPENDIX A: AspenTech Training . . . 233
APPENDIX B: AspenTech and Feedstock Projection . . . 237
APPENDIX C: AspenTech and Exceptions . . . 239
Links in Book . . . 243
References . . . 247

Vendor Acknowledgments and Profiles 257
Author Profile 261
Abbreviations 265

CHAPTER 1

Introduction

Let's begin this exploration of process industries by taking a look into why process industries have the name that they do. Without much thought, we can see that the term relates to the division of manufacturing into different categories. This general definition developed holistically and without much forethought. The actual definition of the word "process" does not describe the industries to which the process is assigned.

A process is a series of actions or steps taken in order to achieve a particular end. It is:

- *A natural or involuntary series of changes: the aging process.*
- *A systematic series of mechanized or chemical operations that are performed in order to produce or manufacture something: the modern block printer needs to accommodate all the traditional plant processes in one shop.*
- *A series of interdependent operations carried out by computer.* — Apple Dictionary

Certainly any manufacturing environment would meet the above definition. Regardless of the type, manufacturing is the "process" of converting inputs to outputs. However, for whatever reason the term "process industries" was used to describe manufacturing of a particular type.

What Is Process Industry Manufacturing?

Process industry manufacturing is the type of manufacturing environment that manufactures products that cannot be disassembled to their components. The products manufactured by the other types of manufacturing environments—discrete and continuous—can be disassembled. In contrast, the finished product of process manufacturing is not simply the assembly of the input products, but is in fact an altogether transformed item. For example:

1. One cannot unthread fabric to get to the original spools and restock the thread as inventory.
2. Once crude oil is cracked into jet fuel, kerosene, tar, etc., it cannot be reconstituted back into crude oil.
3. In some types of process industry manufacturing, the input product is literally "gone." For instance, when coal is converted into electrical energy, it cannot be converted back into coal.

Another common feature of process industry manufacturing is the number of input products versus the number of output products, which is nicely explained in the quotation below:

> *"In a discrete environment, many parts are used to make subassemblies, which in turn are put together to make the final product. In other words, many small parts come together to make a product. For example, it takes fifty thousand or more parts to create an automobile.*
>
> *"Process manufacturing is the reverse. A few products or ingredients result in thousands of different products. In plastic manufacturing, ethylene may result in many different products depending upon how it is processed. In semiconductors, a single wafer may result in many different final parts and in food manufacturing, a single ingredient like oranges could result in many products like fresh juice, oranges, dried oranges, frozen juice in different packages and so on."* — Supply Chain Planning for the Process Industry

Discrete and repetitive manufacturing are more precise than process manufacturing, and the inputs and outputs have less variability, as is explained in the following quotation:

> *"The basic materials utilized by a discrete manufacturing company differ significantly from that of the process manufacturer. The discrete manufacturer "has it easy." When a discrete manufacturer specifies material (parts) to his vendor, he gives a drawing number and very tight specifications (plus or minus a thousandth of an inch, for example) and exact delivery dates. In the process world, materials are defined more broadly, often with wide specification ranges (water content from 3 to 19%, for example) and date ranges due to seasonality or transportation uncertainty."*[1]

[1] Process ERP Partners LLC. *What Makes Process Process?* South Africa Production & Inventory Control Society, 1998.

Additionally, products are significantly more complex in the process industry. An excellent example of this is also provided by the article *What Makes Process Process?*

Basic Material Definitions

Discrete	Simple Process	Complex
Item Number	Item Number	Item Number
Location	Location	Location
Quantity	Quantity	Quantity
	Unit of measure	Unit of measure
	Tank/silo	Catch weight
	Lot number	Tank/silo
	Aging information	Lot number
	Classification	Sub-lot number
		Container ID
		Container type
		Sub-lot number
		Container ID
		Aging information
		Classification
		Quality
		Hazards Information

This shows the extra dimensions to process industry products.

The most common process industry products include the following:

1. Chemicals
2. Metals
3. Petroleum Products
4. Beverages
5. Pharmaceuticals
6. Food
7. Textiles
8. Paper
9. Paint
10. Glass
11. Power Generation (although power generation creates energy, rather than a product, it is actually a form of process manufacturing)

The products listed above are really quite common. They are products that each of us use every day. However, unless they happen to work in the industry, process industry manufacturing resides as a concept in the recesses of people's brains, even though process industries are also one of the most interesting areas of manufacturing to study. Sometimes process industries are interesting because the process is so extreme, as in the case of the blast furnace—a mechanism that amounts to hell on earth and runs in virtual perpetuity. Sometimes they are interesting because the process is so amazingly and tightly integrated and automated that it almost seems to run itself. A good example of such a process is the textile loom. Our present standard of living and the electrified environment in which we live (with seemingly unlimited on-demand power) are due in large part to the extreme temperatures, large size, high-speed rotation, and continuous operation of steam turbines, which have been customized with the most durable and long-lived materials in order to endure the temperatures necessary for high-efficiency conversion of coal or petroleum into energy. While the baked goods we eat may look as if they were baked in a bakery, in fact much of it was produced on a continuous high-speed production line. In fact, most of the final products of the process industry are completely unrecognizable from their input products. This book focuses on how to apply software to the planning problems of process industry manufacturing, but along the way I think you will find that the individual peculiarities of each of the different sub-categories of process industry manufacturing will hold your attention.

As you have determined by now, process manufacturing is one of the three major categories of manufacturing, along with discrete and continuous manufacturing. It is also the least emphasized of the different manufacturing categories and its terminology is less familiar than that used in discrete or continuous manufacturing. After analyzing this issue, it's unclear to me why this is, but it may be because products that are manufactured in the process industry are not as prominent in everyday life versus discrete manufactured products. However, while this is logical for the general population (who thinks about the gasoline in the tank as much as the automobile, or the water that was purified before arriving at the faucet as much as the new iPhone?), however, it is really not logical at all for this predisposition to persist into industry and into software development, which it most certainly does. If the term redheaded stepchild ever applied to anything, it applies to process industry manufacturing. The differences begin right at the terminology used to describe each area. Instead of parts, a bill of materials and a routing (common nomenclature in discrete and continuous manufacturing), process industry manufacturing has ingredients, recipes and a formula. While a background in mechanical engineering would be helpful in understanding discrete and continuous manufacturing, the most appropriate training for process industry manufacturing is certainly chemical engineering.

Many people who work around discrete manufacturing don't actually understand what makes discrete or repetitive manufacturing different from process manufacturing. And these are people who you think should know better. I was amazed when a director at Deloitte once told me that cheese and whey processing was "somewhat discrete." At the time I responded that the company in question was completely in the process manufacturing category because cheese manufacturing involved the processing of liquids, used vats, and had lagging operations.

(That is not start-finish or start-start as would be the case between operations of manufacturing pens or toys.)

I have synthetized much of the information about process industries in this book—at least at a reasonable level of detail— so no one who reads it could make the same mistake as that Deloitte director. As you read this book, some company somewhere is struggling through the implementation of production planning software that their executive decision makers did not know was never designed for the process industry environment. Part of the trick of understanding process industries is simply finding the right analogy. I have selected the easiest applications that I know of in which to demonstrate production modeling, and I have a great deal of experience analyzing production planning systems in detail, thus understanding how software is used to plan all of these different process industries should be a very achievable goal.

Books and Other Publications on the Process Industry

As with all my books, I performed a comprehensive literature review before I began writing. One of my favorite quotations about research is from the highly respected RAND Corporation, a "think tank" based in sunny Santa Monica, California. They are located not far from where I grew up. On my lost surfing weekends during high school, I used to walk right by their offices with my friend—at that time having no idea of the institution's historical significance. RAND's *Standards for High Quality Research and Analysis* publication makes the following statement about how its research references other work.

> *"A high-quality study cannot be done in intellectual isolation: It necessarily builds on and contributes to a body of research and analysis. The relationships between a given study and its predecessors should be rich and explicit. The study team's understanding of past research should be evident in many aspects of its work, from the way in which the problem is formulated and approached to the discussion of the findings and their implications. The team should take particular care to explain the ways in which its study agrees, disagrees, or otherwise differs importantly from previous studies. Failure to demonstrate an*

understanding of previous research lowers the perceived quality of a study, despite any other good characteristics it may possess."

Most books on the topic of process industry manufacturing deal with the business process. There are extremely few books that are software focused that address process industry manufacturing and no books that address specifically planning, ERP, MES or process control for process industry manufacturing.

There are two categories of books that cover planning with software or planning in the process industries. The first category includes books that cover process manufacturing from the technical modeling perspective. These books include a good deal of mathematics or chemistry. They are written by PhDs and written for either PhDs or software developers. Some of these books are listed below:

1. *Advanced Planning and Scheduling Solutions in Process Industry:* This is a very technical book for those who want to delve deeply into the different approaches of optimization in the process industry. For instance, topics include when to use a linear program or a nonlinear program, etc. The author includes the mathematics for the various optimization approaches and this book is targeted toward those with a PhD in the subjects covered.
2. *An Advanced Production Planning and Scheduling System: For Batch Process Industry:* The research that is the topic of this book focuses on batch (as opposed to continuous processing) in the process industry.
3. *Meta Heuristics for Large Scale Process Scheduling:* This book eschews the common approach laid out in books about optimization for process scheduling, in favor of simpler meta heuristics. This book is aimed at PhDs in the field.
4. *Mixing In the Process Industries:* This is a technical book on how to perform mixing operations in a variety of process industries. There are actually several books on mixing for the process industry, but I have included just one example here. Many of these books are borderline chemistry texts.
5. *Supply Chain Planning for the Process Industry:* I reference this book on several occasions as it does a nice job of explaining some generalities about what separates process industry manufacturing from other types of manufacturing.

A second category of books is the general management book. These books typically explain how to implement an improvement in process industry manufacturing. When one compares the number of books like this that are written for process industry manufacturing versus discrete manufacturing, as I alluded to earlier, the comparison is quite amazing. The only books I was able to find were the following:

1. *Lean for the Process Industries:* How to apply lean techniques to the process industry.
2. *Liquid Lean: Developing Lean Culture in the Process Industry:* How to apply lean techniques to the liquid process industry.

I found it curious that both of the management books I found were related to the lean trend. Why anyone would want to apply lean-manufacturing principles to process industry manufacturing is beyond me, and I think you might agree by the time you finish this book.

There is a third category of books that are not related specifically to planning but that cover each specific area of manufacturing within the process industry. For instance, if one is looking for a book on how the production process works in petroleum refining, pharmaceutical production, fermented beverage production, milk production, and so on, there are books available in each of these areas. Furthermore, there are many books available on the manufacturing of the specific item produced. There are books on coffee, yeast, milk, whiskey, and pretty much any other process industry product that you can imagine. If you want to know how to produce a specific item, there is probably a book on the topic. (Although strangely I could find very few books on the paint production process; for such a large industry there is an amazing shortage of published books on the topic.) There are even books like *The Art of Fermentation,* that cover the process industry sub-process of anaerobic metabolism, which applies to a wide variety of products ranging from alcohol to vitamin C and sauerkraut. This book doubles as a cookbook and a book on a specific process, which is performed both in one's kitchen and on an industrial scale. Typically these books are quite technical, as much of process industry manufacturing is essentially the control of chemical reactions in materials as they flow through a plant. In fact, the many biology books I had read throughout the years were good practice for these books, as they are part biology books and part books on a manufacturing process.

All of these books are excellent background to process industry manufacturing. In fact, there is little doubt that process industry manufacturing has the greatest diversity of all the manufacturing types. At the end of the day, process industry planning is about planning the manufacturing of a specific process industry product and each product is a bit different. I read several of these books and each one reinforced the fact that the process industry has the widest variety of manufacturing requirements of any of the manufacturing categories.

Process industry manufacturing has the most specific needs, and it is an industry (or group of industries) that therefore has a complex "code" that by and large has not been cracked by commercially available manufacturing software. This is the primary reason that process industry manufacturing has the highest percentage of custom software developed for it. It is also the reason that many applications that may have worked perfectly well in a discrete production environment, fail in a process industry environment. I was intrigued by the fact that the process industry has no book that covers either the overall process industry for manufacturing or for software. It is granted that there are differences among all the process industries—but unless these companies intend to custom develop everything, they all do have to choose from a limited set of off the shelf applications and there are consistencies and functionality to look for. For instance, they all require recipe management, they all need ERP systems actually designed for their requirements, etc. So at some point, having a completely parochial set of literature is counterproductive because it is not reasonable to expect to find a software vendor that offers just chemical planning or just an ERP system specifically designed for petroleum refining.

While each process industry subcategory has unique qualities, there are also consistencies between the various sub-categories. A few examples are as follows:

1. A large percentage of process industries work with either liquids or gasses.
2. Timing and temperature are critical during the production process across most process industry sub-categories.

3. Frequently process manufacturing has leading and lagging operations.

In fact this book describes many of the similarities across the process industries. Software that works for process manufacturing (and there are very few production planning applications that work for process manufacturing) has functionality that addresses these specific similarities. When one reads through the books on specific process industry product production, it becomes clear that process industries are simply different in their manufacturing needs.[2] However, most business manufacturing trends tend to originate from discrete manufacturing, and process industry manufacturing is repeatedly expected to implement trends that are inappropriate for their needs (see the lean manufacturing books that I referred to previously). It's hard for me to fathom how such experienced people in manufacturing can repeatedly over-generalize from discrete manufacturing to all other manufacturing environments. This book you are reading is one of the few books to address this issue, which is covered in Chapter 3: "The Discrete Manufacturing Focus of Manufacturing Software."

One can find reasonably good academic articles on specific types of process industry manufacturing planning, but the better articles are related to a specific process industry, such as milk processing, petroleum refining, etc. But none of the academic papers that I reviewed were particularly applicable to this book. So while some of my books rely quite heavily on academic literature, this book will not.

A Unique Book

This book has a number of firsts. It is the first book to focus on manufacturing software as a category. It is the first book to analyze the fit between the process industry and the software available. It is also the first book to approach the major sub-categories of manufacturing software; ERP, planning, recipe management, MES and process control. Literally any process industry manufacturing company should be able to greatly benefit from this book. The only thing this book

[2] Actually, process industry manufacturing has a wide variety of types simply within this category. There are simple process industry manufacturing companies that have requirements for lot control; bottling or paint would be a good example. On the other side of the continuum are process industries like meats that have requirements to manage co-products, recycle streams (to process more than once through a resource), multiple formulas as well as characteristics. This concept of either simple or complex process industry requirements was explained well in the paper *What Makes Process Process?*

does not cover—is how these manufacturing applications are integrated to one another. This topic, which is highly relevant and for which there is a shortage of available information was deliberately left out, because it is not specific for the process industry.

The Book's Grounding in Reality

A number of books are interested in perpetuating an unrealistic viewpoint or representation of the actual practice. Management books are probably the worst in this regard, talking about initiatives in companies as if they are going to completely transform the area. However, software books are equally guilty, often making it seem as if change is just as easy as buying and implementing the software profiled—software that of course meets all requirements, with functionality that works every time at every place of implementation. It's interesting to read books about software with which I have personally worked, and see the author make no mention of the problems that I have witnessed numerous times. That type of approach may help sell books, but it doesn't do anyone any service besides the author and of course the software vendor. All of the books at SCM Focus Press take the unusual and independent approach of attempting to reflect the actual reality of software enterprise implementations.

Use of Graphics and Video in the Book

This book is quite similar to the previous books I have written in that there are a number of screen shots of applications. However, this is the first book that I have written that also uses pictures. The reason for this is that process manufacturing is so different (depending upon the subcategory) that photos are required in order to understand what is going on in that particular manufacturing process. This is also the first book where I have included links to YouTube videos.[3] This is for the same reason; without seeing the machines in action, something is lost in translation. Some of the most interesting videos, in my opinion, are the ones that show a continuous manufacturing operation. The degree of coordination and speed is almost unsettling. I have been on many plant tours, wearing glasses and feeling the heat and noise, and I can say that observing the process from the comfort of

[3] One of the downsides of this is that YouTube videos can be taken down at any time. Therefore, if the link that I used does not work, try typing in the description that I have included below the link.

a desk and without other distractions of the environment is an excellent way to learn about various manufacturing processes.

Use of Vendors and Specific Applications in the Book

Like most SCM Focus Press books, this book uses specific examples and screen shots from different vendors. Unlike most books about software, this book showcases more than one vendor. Focusing an entire book on a single software application is beneficial for those who want to use solely the application in question. However, only the biggest vendors like Oracle and SAP tend to be profiled in books that showcase their applications, meaning that only the worst functionality in the software category gets profiled. These books give exposure to companies on the basis of them being big vendors, rather than whether the applications actually showcase good functionality, usability, implement-ability and maintainability. However, I don't want to showcase the worst software in a category; I want to showcase the best.

Many of the people who read this book will probably never get an opportunity to use the applications that I am showcasing. However, the principle at work is to show how a desired functionality can be modeled in software. This approach also serves to ground the book by showing that the capabilities I am describing are not hypothetical.

Areas the Book Does Not Cover

In addition to explaining what I will cover, I want to be explicit about what I will not cover. I thought of covering each of the following areas, but decided the book would be better and meet the needs and interests of more readers if I did not make them a focus of the book.

Production Planning Methods

Frequently, the planning methods employed are a major topic of interest in any software book with a supply chain orientation. In production planning the dominant methods are heuristics and optimization. Planning methods require a good deal of explanation. For instance, even within the method of optimization, there are several different types of optimization. I have written three books that address optimization in one or more chapters, and one book that is related purely to supply planning optimization *(Inventory Optimization and Multi Echelon Planning Software).* I've written two books that are a combination of supply and production methods (*Supply Planning with MRP, DRP and APS Software*, and *Constrained Supply and Production Planning in SAP APO*). There are of course many other books on supply chain planning optimization. I have also written a number of articles on optimization, particularly about the practical usage of optimization in supply and production planning. However, while there are specialized requirements in the process industry for optimization, in this book I focus primarily on the use of commercially available software.

In this book I focus on the areas of production, which are unique to the process industry and how to address those requirements with software. There is so much to cover in the process industry that is unique ***to*** the process industry, and I did not want to divert attention to areas of production functionality, which apply to all industry types.

In terms of ERP systems, one can perform planning in an ERP system, but not very well. In order to access better functionality; it is desirable to use an external planning system (also called an APS or advanced planning and scheduling system). To enable them to be used with process industries, most ERP systems have some type of process modality in which they operate. In SAP ERP this is referred to as the PP-PI or production planning-process industry. Because it is necessary to pass planned orders to the ERP system for processing, I include a chapter that explains how an external planning system is connected and interoperates with the ERP system, using SAP ERP as an example.

Custom Solutions

Process industry manufacturing has the highest percentage of custom solutions of any of the manufacturing categories. However, I chose not to cover this topic as doing so would result in a very technical book. Since I am not a developer or PhD in operations research or related field, I am not qualified to write such a book. I considered adding a case study of a custom-coded solution, but I also think the topic is so involved that it really should be its own book.

It is certainly true that there are many areas of process industry manufacturing for which there are no commercially available solutions. This will not be the best book for people who are searching for information on how to build a custom solution. Instead this book is directed toward those who are looking toward a new commercial software implementation, or who are trying to improve the operations of a current system. Of course, it is also targeted at those who simply want to understand process industry manufacturing better and who will take the software explanations and screen shots as a general explanation on modeling. For people looking for a book on how to integrate ERP, planning, MES and process control software I have a different book that addresses these issues entitled, *Integrating Manufacturing Software: ERP, Planning, BOM/Recipe, MES & Process Control.* This book includes both software information and business process information, so it should be useful to both software types and business types.

How Writing Bias Is Controlled at SCM Focus and SCM Focus Press

Bias is a serious problem in the enterprise software field. Large vendors receive uncritical coverage of their products, and large consulting companies recommend the large vendors that have the resources to hire and pay consultants rather than the vendors with the best software for the client's needs.

At SCM Focus, we have yet to financially benefit from a company's decision to buy an application showcased in print, either in a book or on the SCM Focus website. This may change in the future as SCM Focus grows—but we have been writing with a strong viewpoint for years without coming into any conflicts of interest. SCM Focus has the most stringent rules related to controlling bias and restricting commercial influence of any information provider. These "writing rules" are provided in the following link:

http://www.scmfocus.com/writing-rules/

If other information providers followed these rules, we would be able to learn about software without being required to perform our own research and testing for every topic.

Information about enterprise supply chain planning software can be found on the Internet, but this information is primarily promotional or written at such a high level that none of the important details or limitations of the application are exposed; this is true of books as well. When only one enterprise software application is covered in a book, one will find that the application works perfectly; the application operates as expected and there are no problems during the implementation to bring the application live. This is all quite amazing and quite different from my experience of implementing enterprise software. However, it is very difficult to make a living by providing objective information about enterprise supply chain software, especially as it means being critical at some point. I once remarked to a friend that SCM Focus had very little competition in providing untarnished information on this software category, and he said, "Of course, there is no money in it."

The Approach to the Book

By writing this book, I wanted to help people get exactly the information they need without having to read a lengthy volume. The approach to the book is essentially the same as to my previous books, and in writing this book I followed the same principles.

1. **Be direct and concise.** There is very little theory in this book and the math that I cover is simple. This book is focused on software and for most users and implementers of the software the most important thing to understand is conceptually what the software is doing.

2. **Based on project experience.** Nothing in the book is hypothetical; I have worked with it or tested it on an actual project. My project experience has led to my understanding a number of things that are not covered in typical supply planning books. In this book, I pass on this understanding to you.

3. **Saturate the book with graphics.** Roughly two-thirds of a human's sensory input is visual, and books that do not use graphics—especially educational and training books such as this one—can fall short of their purpose. Graphics have also been used consistently and extensively on the SCM Focus website. Before writing this book, I spent some time reviewing what has already been published on the subject.

The SCM Focus Site

As I am also the author of the SCM Focus site, http://www.scmfocus.com, the site and the book share a number of concepts and graphics. Furthermore, this book contains many links to articles on the site, which provide more detail on specific subjects.

Use of Quotations

This book makes frequent use of quotations, which were recorded during the research phase of the book writing process when I found many insightful comments from several sources. Some of my favorite quotations came from the documents produced by Arena Solutions, and these quotes were very effective at providing additional information about aspects with which I did not have direct project experience. A guide for authors that I had previously read stated that one should limit the usage of quotations and try to rephrase the author's comments into one's own writing whenever possible. I believe they teach this approach in some college writing courses. I have to say I completely disagree with this sentiment. An effective document in a technical field triangulates with the other existing material. This is not to say that one must agree with the quotation and in fact I list several quotations, which I think are out of date given recent software developments. (This should not be seen as criticism of the authors. The authors were writing before some of the software described in this book existed.) I simply do not feel that paraphrasing quotations is the correct way to go about writing the book. I have always been in favor of using quotations and I use them liberally on the SCM Focus website. In fact, I frequently add a quotation after a post is already published because I find a new spin or unique insight on the topic from an author. I also recall teaching supply chain for a year, and how I was irked at reading student papers that did not include a quotation where one should have been included; that is, some sentences came across as minor rewrites of what was obviously someone else's idea. I also have a problem with white papers placed on the web as promotional literature by software or consulting companies that not only lack quotations, but footnotes as well. Some of the best-known consulting companies leave footnotes out of their white papers. This creates the illusion of innovation where none exists. I was once told by Andersen Consulting to not mention that I had found an innovative way to handle an inventory management problem from a book by George Plossl. Andersen Consulting wanted to leave the impression that I had come up with the idea on my own.

Intended Audience

This book is targeted toward people who want to better understand process industry manufacturing the major sub-categories of manufacturing software and how to build effective solutions from commercial software. If you have any questions or comments on the book, please e-mail me at shaunsnapp@scmfocus.com.

Abbreviations

A listing of all abbreviations used throughout the book is provided at the end of the book.

Glossary

The following glossary of terms will be helpful before reading Chapter 3: "The Discrete Manufacturing Focus of Most Manufacturing Software." Some have been taken from PlanetTogether literature. Breeze through these terms now, but return to this glossary when the terms are first used.

1. *Plants:* Correspond to separate physical factories (I use the term "plant" and "factory" interchangeably).
2. *Resources:* Usually correspond to machines, people and tools or subcontractors. They are used temporarily to perform an operation; a resource is a capital asset that is available and used to perform work. Materials are not resources since they are "consumed" rather than used temporarily. In this broad definition of a resource, Galaxy APS considers any machine, tool, die, mold, or even skilled laborer as a resource. The general rule is that if it is a production asset that requires scheduling, it will be defined as a resource inside Galaxy APS. Each operation in a job has one or more resource requirements, each of which specifies the need for another additional resource in the production process. Each resource belongs to exactly one department.
3. *Routing:* A combination of resources in sequence.
4. *Alternate Path:* An alternate path specifies the precedence relationships between operations thus indicating the path that is followed through the

shop to produce a product. Each manufacturing order (M.O.) specifies one or more alternate paths and exactly one default path. When the M.O. is scheduled, the default path is used. A path can converge and diverge, meaning that one operation can have multiple successors and multiple predecessors.

5. *Capabilities:* This is the work that a resource can do. A capability is a specification of a skill or function that can be performed by a resource. For example, a capability might be "cutting" or "inspection." Each resource has one or more capabilities that indicate the types of work it can perform. In addition, each job operation resource requirement has one or more required capability indicating the capabilities that a resource must have to be considered eligible to perform the operation. If an operation, for example, requires a CNC machine and a CNC operator, then you would create two capabilities such as: CNC and CNC-Operator. Then the operation would have two resource requirements: one specifying CNC and one specifying CNC-Operator.

6. *Manufacturing Order:* What is created in order to plan and schedule actual production. A Manufacturing Order (M.O.) is a request to manufacture a specific quantity of a specific product. Each job contains one or more manufacturing orders. Each manufacturing order contains one or more operations and one or more alternate paths that describe the sequence in which operations should be performed.

7. *Operation:* This is the work performed on a resource. An operation is a definition of a single processing step in a manufacturing order. The operation specifies values such as the Cycle Time, Setup Time, expected Yield, Resource Requirements, Material Requirements, etc. Each manufacturing order contains one or more operations. Operations can be "connected" to establish predecessor/successor relationships using the alternate path of the manufacturing order.[4]

8. *Drum:* Used in the Drum-Buffer-Rope metaphor of the Theory of Constraints where the drum is the constraint, the buffer is the material release duration, and the rope is when the material is released.[5]

---

[4] From the PlanetTogether Glossary

[5] Galaxy APS can use Drum-Buffer-Rope scheduling if configured to do so.

Corrections

Corrections and updates, as well as reader comments, can be viewed in the comment section of this book's web page. Also if you have comments or questions please add them to the following link:

http://www.scmfocus.com/scmfocuspress/production-books/process-industry-planning/

CHAPTER 2

Continuous versus Batch Processing

In process industries, one common and quite logical approach to segmenting process manufacturing is by whether it is a continuous or batch process, and so this is a good place to begin this book. Both continuous and batch manufacturing processes apply to products that cannot be disassembled into its original components, however, they differ with respect to the length of the production run and therefore the frequency of changeovers.

Production Planning Categorization

Products Can be Disassembled Into Their Components	Length of Production Run: Short	Length of Production Run: Long
Yes	Discrete	Repetitive
No	Batch	Continuous

When one analyzes the manufacturing categories, it's apparent that process industry manufacturing is divided into batch and continuous, and that these categories correlate to discrete and repetitive in terms of their constancy of production. The differentiation is whether the finished good can be disassembled after production into its component parts.

Interestingly, while the term "process" is used as a category for both batch and continuous manufacturing, there is no such umbrella term for discrete and repetitive manufacturing. This terminology is confusing to many people; poor terminology obscures the fact that continuous manufacturing is similar to repetitive manufacturing, something that would be obvious to people if the terminology had been developed more clearly from the beginning. One of the problems in developing terminology is that while most process industries deal with liquids and gasses, not all do. For instance, textile industries work with solids, but are in the process industry category because textiles cannot be disassembled after production is complete.

Essentially manufacturing environments are segmented along two dimensions:

1. Whether the finished good can be disassembled into its component parts.
2. Whether the processing takes place in a continuous fashion or is in batch.

Production Planning Categorization

	Short	Long
Disasseble-able	Discrete	Repetitive
Not Disassemble-able	Batch	Continuous

One could imagine a renaming of the different manufacturing categories.

The manufacturing categories could be referred to this way:

1. Non-Disassemble-able / Short: Batch
2. Non-Disassemble-able / Long: Continuous
3. Disassemble-able / Short: Discrete
4. Disassemble-able / Long: Repetitive

In fact one could extend the category to include very small production quantities as I have done below:

Production Planning Categorization

	Single Item Production	Short	Long
Disasseble-able	Job Shop	Discrete	Repetitive
Not Disassemble-able	Small Batch	Batch	Continuous

Some people may say that this is just semantics and is therefore a minor issue, but I disagree.[6] Inconsistent terminology negatively impacts learning. Inaccurate terminology has reduced intelligibility and understanding on a number of my projects. A perfect example of this is the misuse of the term MRP and MPS, as described in the following articles.

[6] In fact, in my consulting experience, whenever someone uses the term "semantics," with me, it is typically because a person has used a word incorrectly. The person would prefer not to admit that he or she used the word incorrectly, and so falls back on the statement: "This is just semantics." "Semantics" simply means the study of the meaning of words and symbols. The words "only semantics" should be used when the two words mean the same thing or close to the same thing, not when they mean different things, and should not be used to excuse the use of inaccurate terminology.

http://www.scmfocus.com/supplyplanning/2011/10/02/the-four-factors-that-make-up-the-master-production-schedule/

http://www.scmfocus.com/supplyplanning/2012/07/25/the-misuse-and-overuse-of-the-term-mrp/

Terminology should be immediately descriptive of what that thing is, but also allow the learner to immediately compare that term with other terms. In fact this exercise brings up the question as to why the term "process" exists and why there is no parallel term to serve as an umbrella for discrete and repetitive. Sloppy terminology leads to slopping thinking and interferes with learning.

Process Industry Manufacturing Sub-Categories

The rest of this chapter will describe many sub-categories of process industry manufacturing environments. This book is completely focused on process industry manufacturing, but is not focused on any sub-category or specific type of process industry area. The primary concept behind this book is that it be useful to all the process industry sub-categories and that it help individuals in companies make better software decisions by helping them select and utilize applications designed specifically for process industry manufacturing environments. However, many readers will only work in one of the following sub-categories, so if you wish to only read the section that covers your particular industry, you can do so, and then move on to the next chapter.

Continuous Production

Continuous process manufacturing is literally just that: a process that is run without stopping. The duration of the continuous process can be quite surprising to those who work outside of this production area. Every day we use energy that is both continuously and batch-manufactured. When we turn on a light switch, that power typically comes from a power plant that is run in a continuous fashion. When we get into our cars, we generate power in batch. Energy from the power plant is produced more efficiently than power from our engine.[7] However, limitations in transmitting and storing the power from the power plant to our automobiles is why we have local batch production of energy in our cars (although technologies may change this in the future).

Process manufacturing offers the efficiencies as described in the quotation below:

> *"Shutting down and starting up many continuous processes typically result in off quality product that must be reprocessed or disposed of. Many tanks, vessels and pipes cannot be left full of materials because of unwanted chemical reactions, settling of suspended materials or crystallization or hardening of materials. Also, cycling temperatures and pressures from starting up and shutting down certain processes (line kilns, blast furnaces, pressure vessels, etc.) may cause metal fatigue or other wear from pressure or thermal cycling."* — Wikipedia

The literature on manufacturing provides little coverage on continuous production techniques. In my experience, continuous manufacturing is one of the least explained methods of manufacturing, even though many manufacturing processes really must be continuous. Let's discuss some specific examples.

[7] This is for a number of reasons. First the steam turbines that are used in power plants are much larger than internal combustion engines. They also spin much faster, operate for more of their time at their maximum thermal efficiency, have higher compression ratios, and can operate at far higher temperatures, making for a more efficient conversion of the fossil or nuclear fuel. The internal combustion gas engine has an efficiency of roughly 30 percent, while steam engines have an efficiency of roughly 43 percent. In fact, the only area where power from the power plant loses out to an internal combustion engine is in the efficiency of transmission, as energy is lost through entropy as it is moved across the transmission line.

Gold Refining

In gold refining, where rock that has been mined is placed into mammoth trucks and taken to the processing center, one million pounds of rock are required to be processed to yield roughly 60 pounds of gold (a ratio of 16,666 pounds to 1).

This Caterpillar truck is used to transport rock. It has a payload of 400,000 pounds. With this quantity of material, the continuous process will be able to do so in the most efficient manner. The refining of a number of metals (platinum, silver, etc.) works the same way; it depends upon a very high scale, which comes about through the combination of giant equipment and continuous manufacturing. This mining truck is loaded by the type of excavator shown in the following image.

Mining operations will have multiple trucks so that the excavator always has a truck to load while the other trucks are in transit to the processing center.

The only thing that these specialized factories do is process rock into gold. Notions of "flexible manufacturing" or changeovers simply do not apply. Some continuous processes run for years as the costs for starting and stopping the process are quite high. With continuous manufacturing processes, the efficiency gained from the process is so high that the emphasis is on keeping the process working continually, meaning that work shifts must be 24/7. Input materials must be in ready supply and kept always available, even if that means carrying significant buffer stocks of material.

The Blast Furnace

Blast furnaces produce metals and are enormous. The technology goes back 2000 years, although the technology was not distributed broadly until the thirteenth century.

This is a blast furnace. The giant pipe that connects to the top of the blast furnace carries coke, limestone flux, and iron ore to the top of the blast furnace where it is dropped into the furnace. The furnace itself is tall and cylindrical, with the bottom of the blast furnace being the hottest and the top being the coolest. Hot air is blasted into the side in order to maintain the combustion. Iron is removed from the bottom of the blast furnace.

The blast furnace maintains the highest efficiency when it is run continuously, and because it does just one thing, there is little need to turn it on or off. In fact, according to the Steelworks website, the blast furnace is typically run continuously for between four and ten years. Overall, process manufacturing tends to run longer than other types of manufacturing and the reason for this is explained in the quotation below:

> *"Facilities at the beginning of the production process tend to be capital intensive. In semiconductors, the Fab facility can run into the hundreds of millions. In chemicals, the bulk of the capital is used to build the facilities that create the products for other processes like blending and packaging.*
>
> *"Because of the need to utilize these assets efficiently at the beginning of the production process, there is a desire to produce large quantities of one product before switching to another product. On the other hand, the market would like the responsiveness associated with smaller production runs."* — Supply Chain Planning for the Process Industry

After speaking to people with many process industry implementations under their belt it was enlightening to listen to their stories regarding how it can be interesting to see people more familiar with "discrete" manufacturing work on models of continuous process industry plants, because they tend to set up the calendars with no work on Saturday and Sunday. They assume shutdowns on the weekend, and they must be disabused of this idea as the plants NEVER shut down unless it's for maintenance.[8] Obviously, these rules do not apply to something like a blast furnace. Instead the rule is to keep the blast furnace in continuous operation, and to make adjustments to the settings of the blast furnace rather than actually turning it on and off.

---

[8] Dave Biddle, Senior Application Consultant, PlanetTogether

Interior of a blast furnace, with the molten iron being poured.

Here is the molten iron from a blast furnace.

http://www.youtube.com/watch?v=64c2mHdzxf8

Search Term: Molten Iron

Petroleum Refining

An oil refinery works in a similar way to a blast furnace. Like a blast furnace, it uses a furnace to change the composition of the raw material. However, instead of metal coming out of the bottom, the heated crude oil is separated in a large, tall process unit called an "atmospheric pipestill," where the crude is separated into various product groups based on the boiling points of the component molecules. With the residuum at the bottom and the light gasses at the top, and a number of different products in between, the pipestill has sidestream draw points for each class of product. These draw points are located where the controlled temperature will result in that product. The main reason for manufacturing

changeovers is eliminated with oil refining because multiple products are made from a continuous process. Here is an exact description of what occurs:

> *"Inside the atmospheric distillation unit are horizontal plates called trays with mushroom shaped bubble caps to encourage the condensing of various products at different temperatures. The lightest products, such as methane, ethane, propane, butane, naphtha and gasoline, rise through the trays to the top of the tower condensing. The medium weight products such as kerosene, jet fuel, diesel and heating oil, condense in the middle. Heavier products, such as residual fuel oils, condense lower down. Finally the heaviest, tar-like products, such as bitumen settle to the bottom of the distillation tower, referred to as the bottom of the barrel. Separating the light tops from heavy bottoms in the atmospheric distillation unit is referred to as topping the crude."* — Oil 101

Oil refining breaks the mold in that it is a continuous process, but it does not require separate manufacturing lines to make different products. The various sidestreams may or may not require subsequent processing such as: removal of sulfur; cracking of larger, longer molecule material to make it suitable for use in gasoline; or conversion of lower quality, less valuable material to higher quality, more valuable product. There is great flexibility in the routing and the processing of the refinery streams to make a thousand different products as needed.[9]

---

[9] Interestingly how production is measured changes in petroleum refining. *"Nothing in a petroleum refinery is done by weight: it's all by volume. There is too much material moved in order to use anything other than volume. Weight measurements are a luxury afforded to pharmaceuticals and consumer goods."* — David Biddle, Senior Application Consultant, PlanetTogether

The input material to an oil refining operation is crude oil. Crude oil is actually a complex combination of either thousands, or in some cases, hundreds of thousands of compounds.[10] From this crude oil the petroleum refining industry makes more than 2,500 products. This principle can be demonstrated by comparing water to crude oil. If one brings water to boil and leaves the temperature constant it will boil all the water out of the vessel—because water—except for trace amounts of minerals dissolved in the water is one molecule with one boiling point. If crude oil is brought to boil at one temperature and that temperature is kept constant, it will boil and then stop—until the temperature is changed. Repeatedly changing the heating temperature such that more of the crude oil vaporizes and is boiled away is called the distillation curve.[11]

What I have explained above is just the first stage of petroleum refining, which is the separation of crude oil into its major elements. However, like the blast furnace petroleum refining runs for years without being brought down for maintenance. A fluid catalytic cracking unit (FCC) will run from two to three years between shutdowns.[12] As with a blast furnace, oil refineries are high volume operations, with most processing between 100,000 and several hundreds of thousands of barrels of crude per day.[13] An oil refinery is essentially a series of interconnected resources that result in a wide variety of finished goods and requires various degrees of processing. As an example, after the crude oil is distilled, one output is diesel oil, which is sent to one other processing resource, a hydroteater where gas is removed and hydrogen is added. At that point the finished good of diesel oil is created. However, to produce the finished good of gasoline, it must first be sent to a gas processing resource, which will then produce fuel gas. This must be sent to an anine treating resource, which will then result in refinery fuel.

[10] Petroleum Refining in Nontechnical Language
[11] Petroleum Refining in Nontechnical Language
[12] Wikipedia
[13] Wikipedia

This is a distillation column; distillation will be discussed in the Brewing, Winemaking and Distilled Beverage section of this chapter.

With separation there is no chemistry actually involved in oil refining; it is merely the separation of the crude oil into its composite parts.[14] Refining also consists of cracking, which involves the use of catalysts for breaking large molecules, reshaping, which changes the configuration of molecules, combining, which makes larger molecules from smaller molecules, and treating—removing contaminants, are, in fact, chemistry. Of course after refining is complete and the finished good is prepared, an entire sub-industry called petrochemicals performs extensive chemistry on these products.

[14] Petroleum Refining in Nontechnical Language

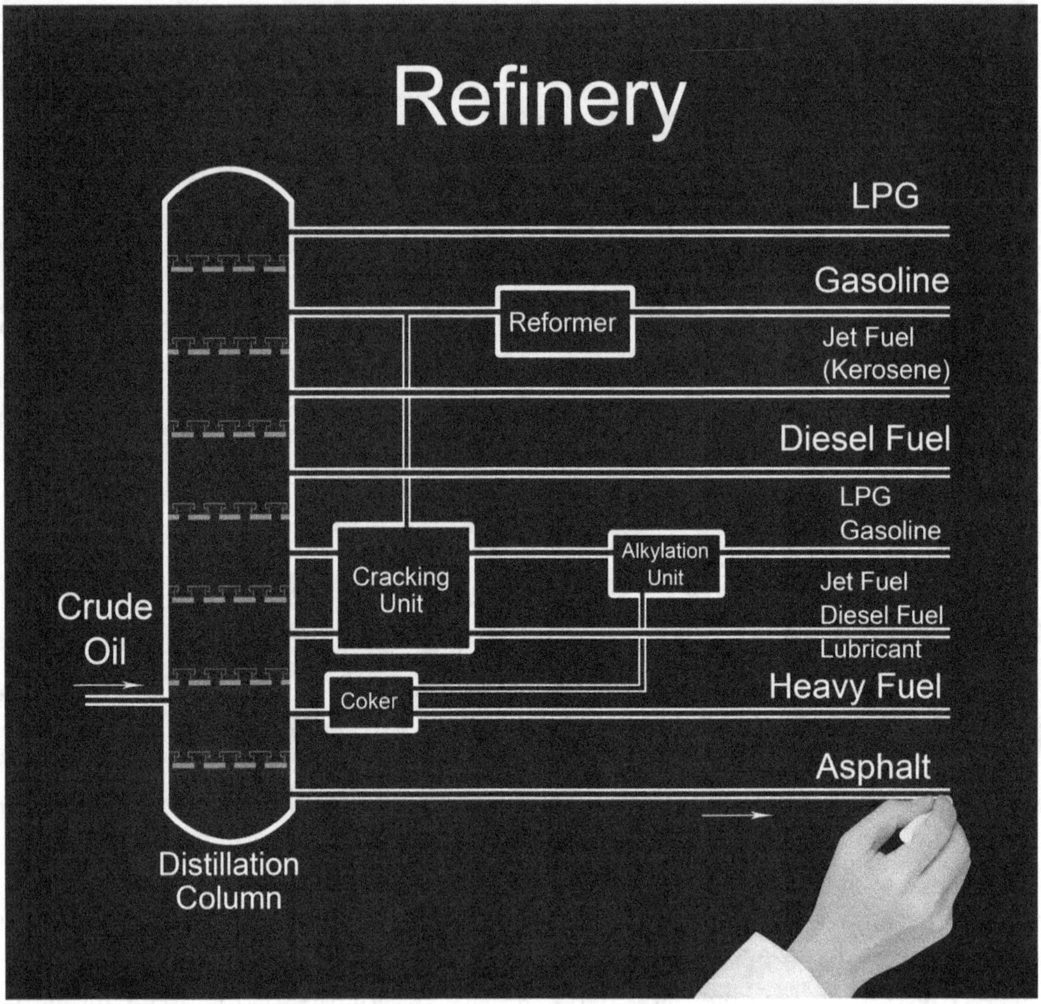

This is a simplified schematic of what occurs in a petroleum refinery.

The following shows how crude is distilled into other products.

http://www.youtube.com/watch?v=2xzYf8IL_FE

Search Term: Oil Refinery

Natural Gas Processing

Natural gas production/processing is classified as a category of petroleum refining, as reservoirs that are primarily natural gas, have dissolved crude oil in them, and reservoirs that are primarily crude oil usually have dissolved natural gas in them. Natural gas can also be found in coal beds, so called coal bed methane.[15] When associated it goes through a field separator and the heaviest drop out in what is called condensate.[16] When natural gas is found in an isolated form from petroleum it is referred to as "unassociated." The four output products of gas processing are methane, propane, butanes and natural gasoline.[17] Methane is the primary constituent of what is generally referred to as natural gas, typically about 87 percent by volume—other constituents being carbon dioxide, nitrogen and hydrogen sulfide.[18]

> *"Natural gas is a complex mixture of hyodrocarbon and nonhydrocarbon constitutents and exists as a gas under atmospheric conditions. Virtually hundreds of different compounds may be present in natural gas in varying amounts. Even two wells producing from the same reservoir may produce gases of different composition as the reservoir is depleted."*[19]

Natural gas processing is a series of continuous processes that will remove water (sometimes crude oil) and impurities and then separate the natural gas into the following each with the following boiling[20] points:

- Propane: –44 degrees Fahrenheit
- Butane: –31 degrees Fahrenheit
- Ethane: –128 degrees Fahrenheit
- Methane: –257 degrees Farenheit

---

[15] Wikipedia

[16] Petroleum Refining in Nontechnical Language

[17] Petroleum Refining in Nontechnical Language

[18] Wikipedia

[19] Handbook of Natural Gas Transmission and Processing

[20] The term "boiling point" is a bit misleading. The definition of a boiling point is simply the temperature at which or the amount of energy at which a liquid molecule will separate from its nearest neighbors to form a gas molecule. (Elmhurst College)

As was discussed in the section on petroleum refining, this is a distillation curve, except the distillation curve for gas is quite low. This is why "gas" is vaporized at room temperature. It is why gas can be liquefied for transportation and storage when taken below its boiling point.

Gas is liquefied for efficient storage, as in these tanks.

Textiles

One often thinks of textiles as simply the assembly of textiles into clothing, automotive seat covers, etc. These processes are labor intensive. However, the step prior to cutting and assembly is the actual creation of the textile material, and this step is in fact quite automated.

This is a massive loom that runs in a continuous fashion. Modern looms are connected to enormous banks of spindles, which serve as a near constant supply of input material as the bank of spindles shows on the following page.

Continuous processes require continual supplies of raw materials. Any shortage of raw materials would mean stopping the process, which would result in significant inefficiencies. Continuous manufacturing processes have the highest efficiency, the lowest per unit cost, but also the highest investments in machinery. Continuous manufacturing is the eventual state when variability has been controlled in the manufacturing process and the process has been fully automated.

These looms are scaled up and more sophisticated versions of the following loom.

Weaving was once a manual activity and in some parts of the world is still done that way. The first powered loom was invented in 1830 in England. Looms have become faster and more efficient ever since, to the point of sophistication where they are continuous operations. A major part of our standard of living is based upon the increased productivity due to the scaling up and electrification of factories. This productivity increase was greater than all other productivity increases before and since—including the electronics and computing revolutions. This is why productivity gains spike so dramatically after 1820—which was the period of the first industrial revolution. Right now, while you are reading this, giant looms are running 24/7, rotating at 1200 RPM, turning out prodigious quantities of textiles at a very low cost. It's hard to appreciate the high speed and continuous nature of these looms until you see one in action. I have included a YouTube link that shows exactly this.

http://www.youtube.com/watch?v=V5e1YKkZFhc

Search Terms: Loom

This video shows the complexity that is involved in the set-up of this machine. Knowing this complexity, it would be hard to make the case that setting up all of the spindles to produce a different product would be a good use of time and the machine. This is a textbook example of continuous manufacturing processes: ***the efficiency of continual use is simply so high it trumps many other factors.***

Paper

Paper and steel are similar in that the raw material manufacturing process in both cases, the front-end process is continuous, but once the raw material has been created, it becomes more of a discrete process of slitting, coating, printing and whatever downstream processes are added to make the paper ready for the customer.[21]

Paper began as a batch manufacturing process, but eventually was converted to a continuous process. The Fourdrinkier machine, shown in the second image on page 45, goes all the way back to 1799. It was responsible for greatly reducing the price of paper, and therefore increasing the affordability of books. Interestingly, in the history of bookmaking, economical paper along with developments in metallurgy and in type casting were foundational to the improvement of the movable type printing press in increasing the quantities of books in circulation and therefore literacy and the ability to share ideas. In most history books the Gutenberg printing press is treated as a solitary invention, when in fact it relied upon both previously reduced paper prices and the improvements in metallurgy as well as

[21] Dave Biddle, Senior Application Consultant, PlanetTogether

Gutenberg's main contribution, the efficient casting of metal letters.[22, 23] Gutenberg himself had a background in metalworking and experimented with the alloys that he would eventually use for the letters of a printing press in his earlier career as a mirror maker. Paper is now considered a commodity and because of such high production efficiency, is an incidental cost in finished goods except possibly for specialty products such as photography books or high-resolution posters and artwork. Because of the Internet, and e-books, paper is not even critical to literacy as it was just a few decades ago. This all makes it easy to forget how central paper was to the development of advanced civilizations.

---

[22] The importance of improved paper manufacturing techniques is explained in the following quotation.

> *"Indeed, when Gutenberg printed his initial Bibles on vellum, they were nearly as expensive as the handwritten copies that they were meant to replace. Absent the new paper technology being developed in Fabriano, the name Gutenberg and his marvelous invention might have disappeared from history."*—Masters of the Word.

In fact, the Gutenberg press was horrendously inefficient by any later standard. In 1450 it took **two years** for Gutenberg to produce *"180 Bibles at 1,282 page each,"*—Masters of the Word. This is a production rate of **7.5 Bibles per month**. Historians estimate that the labor required for this single undertaking to be roughly 6 pressmen and at least 24 typesetters in order to run 6 printing presses for that two-year period of time. If the labor was correctly estimated to be 30 men, then this would come to each man producing ¼ of a Bible in one month of work. The estimated production rate of a single printing press is roughly 12 impressions per hour. A single impression could have made between 2 and 16 double-sided pages. As these were the first presses of their type, it's likely they were toward the lower end of this size and therefore productivity continuum. This meant that these first Gutenberg-produced books were **very expensive**. Depending upon the quality of the book, they could range from one to two times the annual salary of a skilled craftsman. This is a type of expense for what is now quite inexpensive—and this manufacturing limitation meant that being literate was more a function of one's finances than one's intellect.

[23] Printing is a type of manufacturing. Printing with moveable type printing presses preceded Gutenberg by 400 years, but Gutenberg's contribution was *"in devising a method for mass producing type: the type mold process."* — Masters of the Word, as well as the introduction of oil based ink. So the Gutenberg "press" was not only itself allowed for an enhanced manufacturing process, but the input to the book making, or to the use of the press (along with book binding) was the enhancement of the manufacturing processes of two inputs—one being the paper, and the other being the components that made up the Gutenberg printing press itself—the metal letters. However, Gutenberg's ability to mass-produce metal letters was dependent upon advances in metallurgy—which were also recently improved prior to Gutenberg's innovation. The actual press itself was not where the innovation lay, as it had been used for some time for printing prior to Gutenberg. It was simply adapted from a wine or fabric press.

This is a vat of wood pulp. Pulp is liquefied, heated, bleached, agitated, has its PH level adjusted, and treated with chemicals in order to prepare the pulp. At this point, the pulp is between 80 and 85 percent water. A series of tests are performed on the pulp before it is considered appropriate in its characteristics to be processed by the Fourdrinkier machine.

All of the manufacturing steps that are now performed in an automated fashion (as shown in the following images) were at one time in history performed in batch, and the overall process was much slower and, of course, paper was much more expensive, making its finished goods more expensive.

This is the rolling portion of a Fourdrinkier machine. This converts paper pulp into the finished good of paper. The large roll squeezes the water out of the pulp. Later stages of the machine dry the sheet of pulp, where it is smoothed.

Here is the finished product being moved for storage by a specialized paper roll forklift.

http://www.youtube.com/watch?v=jGH7kQ30SKo

Search Term: Paper Manufacturing

Commodity Chemicals

"*The chemical industry products over 70,000 products from acetyls to zirconium chemicals.*"[24] The chemical industry splits very conveniently into several sub-categories that are fine chemicals, specialty chemicals, inorganic chemicals, petrochemicals, pharmaceuticals, renewable energy, and biopolymers.[25] Although,

[24] AspenOne Supply Chain for Chemicals, AspenTech, 2011
[25] Wikipedia

generally speaking, it is most common to simply divide the chemical industry into commodity and specialty chemicals. Common commodity chemicals include broad categories such as acids, waxes, solvents, etc., while the specific chemicals would be things like acetone, ethanol, naphtha, ethylene oxide. Some of the output of oil refining serves as a feedstock for petro-chemicals.

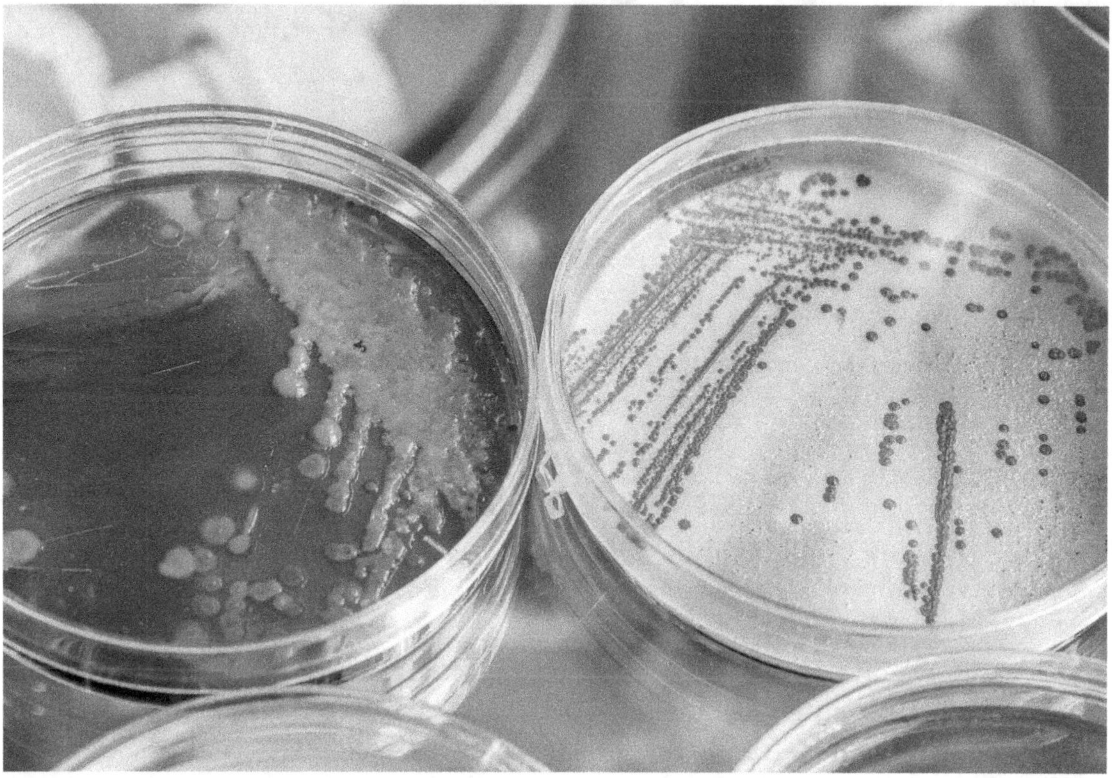

Chemical plants simply engage in large-scale chemistry, the type that is performed in these petri dishes.

As stated there are a huge number of different chemicals that are produced. The image above would be a very common environment for a wide variety of chemical production.

Commodity chemicals are least expensive, and the market for them is extremely large, so not surprisingly, commodity chemicals tend to be manufactured continuously, and specialty chemicals tend to be produced in batch. Actually, it would be difficult to produce commodity chemicals with the necessary margin if a batch process were used. In fact, continuous production and great strides in efficiency have the typical commodity chemical selling for less than 50 cents per pound.[26] Commodity chemicals represent around ⅓ of all dollar output globally of the chemical industry (but of course far more by volume).[27]

---

[26] Wikipedia

[27] Sectors of Chemical Industry

Commodity chemicals are made in a massive scale as with this chemical plant.

Because of the enormous scale and continuous operation of commodity chemicals, and their relatively stable demand, they are far easier to plan than other types of chemical manufacturing operations.

Power Generation

Power generation uses both continuous and batch production. For example, some plants are referred to as base load plants and are in continuous operation. These plant types include nuclear and coal-fired plants. Other plants are called peak power plants. These plants are powered with higher cost energy sources, and include natural gas power plants, which can perform batch power generation—bringing up power when needed and cycling down when it is not.

Nuclear power plants are in continuous production.

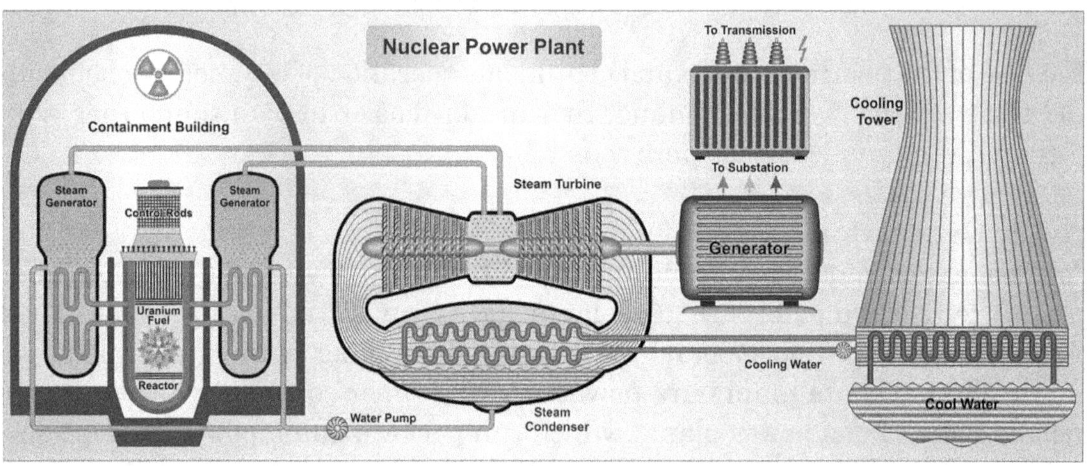

A nuclear power plant works much like most other power generation, only the fuel source is different. However, as with just about everything, it comes down to turning a turbine. The turbine is integrated to a generator that converts mechanical movement/energy into electricity.

http://www.youtube.com/watch?v=MGj_aJz7cTs

Search Term: Nuclear Reactor

This is a steam turbine; the blades have been designed to match with the low density of steam.

http://www.youtube.com/watch?v=AyAd-gLO9CE

Search Term: Steam Turbine

The same principle applies to "windmills," which are actually officially referred to as wind turbines in the energy industry. Windmills were referred to as such because they often used their power to mill grains. However, milling is simply the application of the power, and not related to power generation. The actual electromechanical principle at work is rotation caused by the "turbine" which are the blades—and is converted to electricity. Wind turbines are designed to be run in a continuous fashion, and except for maintenance downtime, there is no advantage to stopping a wind turbine, but of course, oftentimes the wind does not comply—which is why they can never serve as a base power supply.

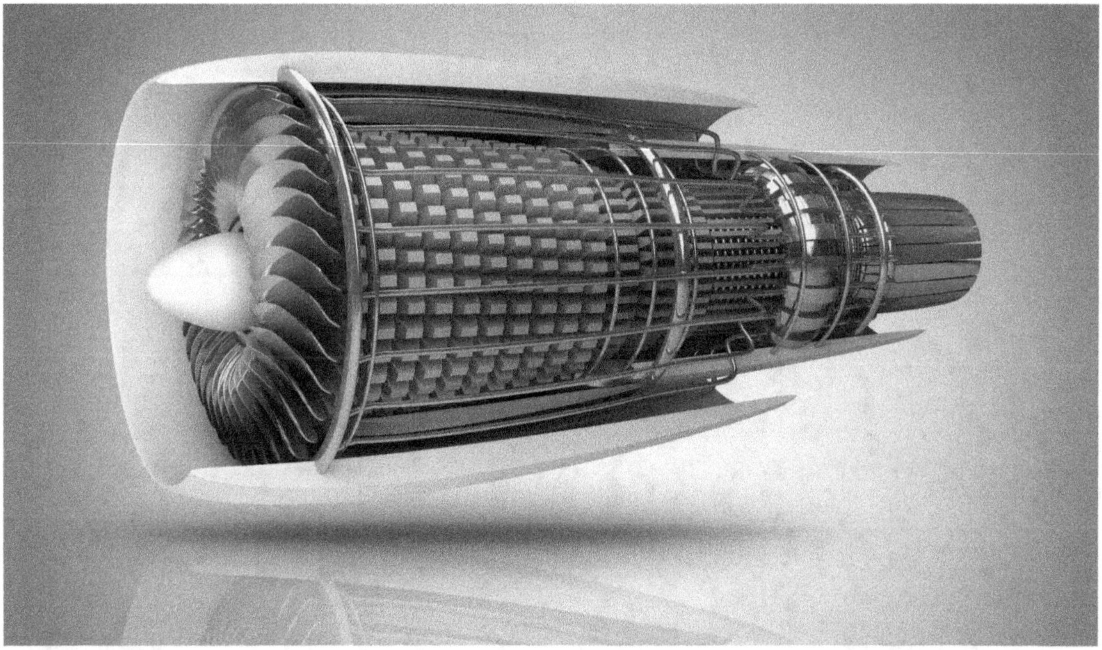

A similar principle applies to the jet engine, also known as a jet turbine. The blades of a turbojet produce air compression before it is sent to a combustion chamber. A turbojet produces continuous power for the length of the trip, but follows the same principles as land based continuous power generation. When we take an airplane trip we rarely think that we are boarding a manufacturing facility, but that is what the jet engines on airplanes actually are.

Dams are by far the largest source of renewable energy on the planet—something that can be surprising considering that waterpower predated not only other renewable or green power generation technologies that often receive much more press coverage, but also all later power generation technologies.

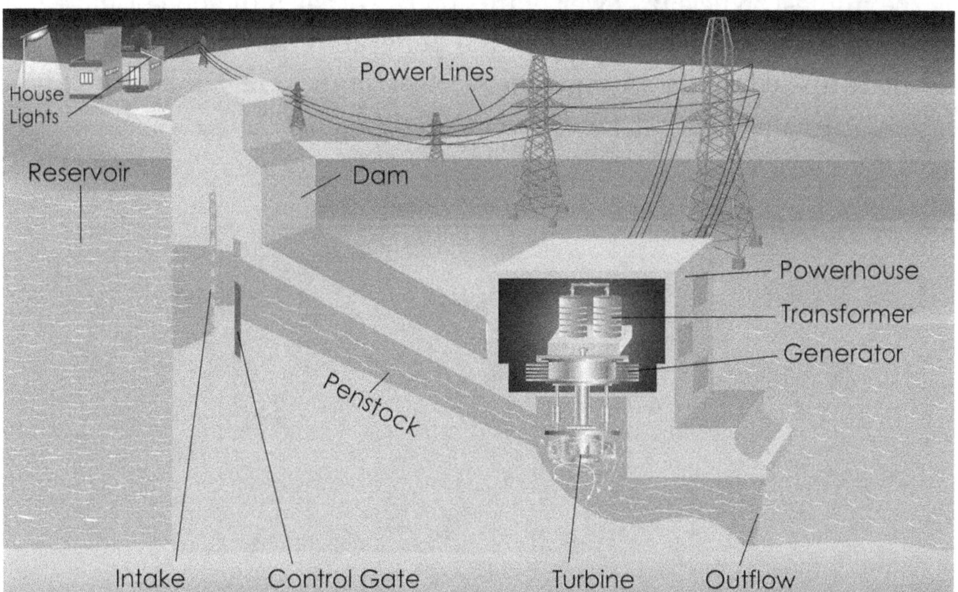

Once again the omnipresent turbine is at the heart of the power generation capability of a dam, but in this case it is a water turbine rather than a steam turbine. The water turbine

turns much more slowly, and the volume of water, which is released from the reservoir, controls its speed. Upon turning, the exact same sequence of generator to transformer to power lines that applies to other forms of power generation applies here as well. This takes mechanical energy—which is a form of non-transportable energy—and converts it into the most transmittable energy—and energy that can be transferred at great distances at very close to the speed of light.

http://www.youtube.com/watch?v=rnPEtwQtmGQ

Search Terms: Hydroelectricity

Dams generally produce continuous power but not at the same rate at all times. Dams **change the amount of water released from the reservoir in order to meet demand**. Dams have another technique of matching the supply of power with the demand by pumping water that has previously been released, back up to the reservoir, called pumped storage hydroelectricity (PSH)—which of course will take more power than the previously released water produced—but is done in order to use off peak energy and allows the dam to sell electricity when the price is the highest. Nuclear power plants on the other hand do not have the same ability to moderate the power level being generated and run most efficiently at a constant rate—which is why nuclear power plants, in addition to their high energy generation capacity, are utilized as "base power" plants.

Coal still produced the majority of power for non-transportation purposes in the vast majority of countries. As with a nuclear power plant, the main objective is to achieve optimal operating efficiency. This means using process control to keep the boilers operating in their optimal range, while keeping the thermal stress factors to a minimum. Some forms of power generation have pollution concerns that many other types of manufacturing do not. Coal fired plants are the most polluting of all the power generation types, and must focus on keeping carbon dioxide and nitrous oxides to a minimum for a set amount of power to be generated. Process control helps to not only maximize power output with lower wear and tear on the power generating equipment, but also to minimize pollutants.

Coal plants not only have to deal with the energy producing resources, but with environmental units, which reduce pollution. One example of this type of resource is an emission control system, for example a gas de-sulphurization unit.[28] Therefore, both of these types of resources must be planned and process controlled in order for the energy produced to be produced on specification. These environmental units can actually be modeled in planning software—by using simulation—to determine the effect of the environmental units on energy generation before they are installed, and can allow the utility to model a number of environmental units to determine which one will work most seamlessly with their power generation units.

Power generation is a type of process industry manufacturing in which there is literally one "SKU" produced—electrons or electricity. Some types of plants need to keep a steady rate of production, while other types of plants can be modulated. The problem of power generation planning is both simpler from a modeling perspective, and more specialized—and vendors like PlanetTogether and SAP do not compete in this area, and while there is some planning, the demand is well known and the capacity is well known, which is often not the case in discrete manufacturing. There are also far fewer resources to manage than in most other forms of manufacturing. If we take the example of a dam, there are typically less than ten water turbines, which are identical. Each type of plant will typically have one input product—water in the case of dams, nuclear rods in the case of a nuclear power plant, and coal for a coal fired plant. This single input material is constantly supplied to the plant so the procurement complexity is low. Rather, this type of planning is a natural fit for a number of AspenTech products, which combines planning with manufacturing execution and process control.

---

[28] Removing sulfur is a constant issue in refining and energy generation. *"One of the annoying aspects of nature endowed on crude oils is the differing amount of sulfur content in various types of crude oil. To complicate this bequest, the sulfur is not in the form of elemental sulfur, a chemical all by itself, but is almost always a sulfur compound. It is chemically bonded to some of the more complicated hydrocarbon molecules so that it is not easily separated from the pure carbon compound. That is not until it is burned. Then it will form one of the several smelly and otherwise environmentally objectionable sulfur/oxygen compounds."*—Petroleum Refining in Nontechnical Language

Producing at less than capacity, or producing in batch, makes very little sense for either nuclear or coal-fired power plants. These are also environments where it is better if lean proponents do not show their face. Nuclear and coal fired plants tend to be setup as base capacity plants—so they are run in a continuous fashion while other types of power plants cover the peak loads. Furthermore, excess energy capacity can typically be sold. Even across thousands of miles, energy can be transported at roughly half the cost of its generation.

Nuclear power plants have very high fixed costs, and these costs must be amortized over continuous production. It simply makes sense to sell energy to the grid if a power plant has excess and vice versa, but of course the demand must make it worth while. Electricity can be transferred at very close to the speed of light; therefore, transmission from any point to any point—as long as the transmission lines exist, can handle the load, and are in good functioning order—can occur essentially instantaneously. Furthermore, because of the extensiveness of the power grid, a buyer can frequently be found as energy can be sold to locations that are across continents. For instance, the state of New York purchases a substantial portion of their overall energy demand from Canada—and with the efficiency in transmission, it is even conceivable that some states could purchase their power from other states rather than having power generation plants of their own.[29]

[29] Canada sells about $4 billion worth of electricity to the US per year. On the other hand the US sold about ½ as much back to Canada—primarily to be used during peak demand.

Power lines of this type, which allow for a power grid, were a tremendous development, allowing not only large scale, and more efficient power generation, but also the development of national and international power grids—which means that power can be efficiently traded between regions—and regions can serve as backups for other regions. Even transfers over thousands of miles of power lines will still only cost roughly half the cost of the initial power generation.

Water Treatment/Water Purification

Water treatment/purification serves multiple purposes. One is to purify water for drinking purposes. Another is to treat wastewater. Common stages in water treatment/purification are the following:

Intensive Treatment Purification

1. *Pre-treatement:* Often removes large debris from the water and allows some natural purification to take place.

2. *pH Adjustment:* Making the water as close to a pH of 7 as is reasonably possible. This can be accomplished by adding lime, soda ash or sodium hydroxide to increase alkalinity or acids to make the water more acidic.
3. *Coagulation and Flocculation:* Remove particles suspended in the water.
4. *Sedimentation:* Allowing suspended particles to settle to the bottom.
5. *Filtration:* Often rapid sand filters are used that moves water through treated sand to remove organic compounds. Membrane filters are also used.
6. *Removal of Ions and Dissolved Substances:* Also known as water softening.
7. *Disinfection:* Removing harmful microorganisms.

Distillation
Discussed in the section on Brewing, Winemaking and Distilled Beverages.

Reverse Osmosis
Applies pressure to push water through a membrane.

Water is treated as a continuous process, and quite under the radar. It is a very important but very low profile sub-category of process industry manufacturing.

http://www.youtube.com/watch?v=SC8QRs77iH0

Search Term: Water Treatment

Steel Rolling

There are two different types of steel rolling—"hot rolling," which is when the temperature of the metal is above its recrystallization, and when it is below this temperature it is called "cold rolling." Cold rolling is actually a misleading term because cold rolling still requires the metal to be heated; it is simply not heated as much as with hot rolling. Hot rolling costs less, but cold rolling is used for metal that requires higher tolerances as part of its specification. In many rolling plants, they are integrated with a blast furnace, so the slabs do not need to be reheated before being rolled or formed—which of course saves energy.

Steel rolling is a type of metal forming—rolling just happens to be an efficient way to form the steel—which is subjected to great pressure. The heated metal combined with the pressure is what allows the forming to be completed. In essence, these are the same two forces that were once applied by blacksmiths, but, of course, at far more efficiency. With the blacksmith, the metal is similarly heated, but the cylinders and other forming presses take the place of the blacksmith's hammer (technically the blacksmith is engaging in forging—a separate metalworking process, while rolling is forming, but the force applies and are conceptually similar). So much of what we rely upon today—from train tracks to steel girders for buildings, to the chassis, axels and bodies of our automobiles—would not be possible without this ability to form steel.

Here is one form of steel—a coil of steel, which is an input product to stamping operations.

http://www.youtube.com/watch?v=6xnKmt_gsLs

Search Term: Steel Rolling

The following images show one of the common characteristics of process industry manufacturing—one is that a single input can result in multiple finished goods. Prior to rolling the same input product can result in...

...Metal pipes...

...railway tracks...

...or steel wire. The different finished good configuration changes the roller configuration. For instance, the well known railway track "H" pattern requires two sets of rollers, one to thin the middle, and one set to flatten the top and bottom. Creating metal plate on the other hand, requires just one set of rollers. Many rolling plants are specialized in just one type of finished good—and are typically named for that finished good—so for instance a plate rolling plant or a track rolling plant.

The Efficiency of Uninterrupted Manufacturing

Continuous and repetitive manufacturing is the most efficient methods of manufacturing. In both manufacturing environments they have very little downtime, and the continuous nature of these manufacturing processes is able to leverage large capital investments into highly automated machinery. Repetitive is the term used to describe uninterrupted manufacturing environments where the finished good can be disassembled into its inputs, while continuous is the term used to describe uninterrupted manufacturing environments where the finished good cannot be disassembled into its inputs.

Continuous manufacturing does not mean that the exact same product is produced. A variety of manufacturing adjustments can be made without performing a changeover that changes the type of material produced. One simple example of this is the temperature on a catalytic cracking unit (used in oil refining) that can be run at higher temperatures to create smaller molecules, but creates more coke and less product.[30] The concept of requiring a changeover in order to produce a new product is really a holdover from discrete and batch manufacturing. On the other hand, in continuous manufacturing, **there can be more than one way to make the same finished product.**[31] One other variable is that the output product not only varies based upon the processing, but also upon the input product. A good example of this is the various types of crude that an oil refining operation can produce. A light, sweet crude (a crude with a low sulfur content—less than .42 percent—and with a higher percentage of molecules that can be converted into gasoline and diesel) would be processed differently than a sour crude, which has high sulfur content.[32] Sour crude requires preprocessing to remove impurities; this process is called "sweetening" the crude, which actually occurs prior to the crude being placed onto tankers so that the vapor pressures are reduced, making it safer to put into tankers. The crude is processed into diesel and fuel oil, but not gasoline.[33] Unsurprisingly, sour crude is more expensive to process. Therefore, the choice of feedstock results in very different output products and requires different processing steps.[34]

Batch Production

In process industries, batch production is where the production process is stopped and restarted in order to produce another batch of the same product or

[30] *"Coke is pure carbon that can be used for fuel."* — Wikipedia

[31] Dave Biddle, PlanetTogether, Senior Application Consultant

[32] Wikipedia

[33] The world is running down its supplies of oil, and the most desirable light sweet crude is less and less available, which means sour oil, which requires more processing, is more commonly used. This mismatch between feedstock and processing ability has left some sour crude without any buyers.

[34] Providing the information that purchasing requires to make the best decision possible regarding feedstock is a main benefit of an effective planning and scheduling application. Simulation allows a company to decide between different possible feedstock options, and this information can be shared with procurement/commodity traders who work for the company and can bring in their knowledge of the market for crude to make the best match between feedstock type and the processing capacities of the oil refinery. This is actually an area of emphasis of a specific product called Aspen PIMS by the software vendor AspenTech. Overall, AspenTech has a very detailed offering for petroleum refining.

of a different product. When batch production is applied outside of the process industry, it is called "discrete manufacturing." Discrete and batch manufacturing are not the shortest duration production run environments. The manufacturing environment with the shortest of all production runs is called "job production," and it was the predominant form of manufacturing processing prior to the first Industrial Revolution. Let's take a look at a few examples of products that are often produced in batch.

Baked Items

An excellent example of batch manufacturing—and one that anyone can relate to, and many of us have performed ourselves—is baked items. Even though baked items are also made in continuous manufacturing, baked items can be frequently made in a batch.

Here a batch of rolls is ready to be removed. The same oven can be used to make another batch of the same rolls or different rolls, or something different altogether.

Clearly these cookies are made in a continuous flow. Some changes may be made; for instance, the filling may be changed or the cookie/biscuit shape altered or the material that the cookie is made of changed, but there are a very limited number of changes in size that could be made to this continuous operation. There is an inverse relationship between efficiency and flexibility. The trays above can make any number of items, but for pure efficiency, nothing can beat a dedicated continuous machine.

Baking is one of several types of batch processing that leverages yeast in its production. Yeast is critical to baking because it converts sugars in the dough into the gas carbon dioxide—causing the dough to rise.[35] These air pockets are responsible for the spongy texture of bread. Two other forms of batch process production, which leverage the many capabilities of yeast, are brewing and winemaking.

[35] Wikipedia

Brewing, Winemaking and Distilled Beverages

Brewing is the process of adding water to prepared cereal grains and fermenting them with yeast.[36] The process is roughly as follows:

1. *Malting:* The process of drying germinated cereal grains into "malt" to get the moisture content down below 14 percent. (Different roasting times determine the malt's colors—the malt color directly relates to the beer color.)
2. *Milling:* Breaking the malt into smaller pieces.
3. *Mashing*: Converting the starches released during malting into sugars to be fermented. Grains and water are mixed to create a cereal mash.
4. *Lautering*: Straining the sugar-rich liquid that results from mashing so that the sugar is extracted from the grains.
5. *Boiling*: The beer "wort" is boiled—along with the addition of hops (used for flavoring, bitterness to balance the sweetness of the malt, promotion of a foamy head and as a preservative) and other flavorings. This is where chemical reactions take place.
6. *Fermenting*: (Explained on the following page)
7. *Filtering*: Not done with all beers, but removes solids and makes the beer clearer, and removes some microorganisms if the filtration is sterile.
8. *Aging*: (self evident)
9. *Packaging*: Placing the beer in cans, bottle, kegs and bulk tanks.

[36] Wikipedia

Brewer's yeast[37]—described in Chapter 4: "Co-Products and By-Products." It is a controlled biological reaction with the cereal grains serving as a food source for the yeast. Brewing, as well as other forms of alcoholic beverage production, involves fermentation followed by distillation in order to purify the intermediate product. The overall brewing process takes roughly 4 weeks.

Where fermentation occurs. After the water, grain, yeast and additives to optimize the yeast's metabolism have been added, optimal fermentation comes down to temperature and

[37] Baker's Yeast—aka saccharomyces cerevisiae is used in the production of wheat-based products. Yeast is actually a fungus and considered the earliest domesticated organism. Fungus is its own kingdom—separate from plants, animals, protists and bacteria. Yeast, as with all fungus do not require sunlight to grow, which is why they have no problem fermenting in closed vats as shown in the image above. They are the primary organisms harnessed for fermentation, which is also used to preserve as well as change the taste of many foods. They convert carbohydrates to carbon dioxide and alcohols. Yeast is used in bread, alcoholic beverages and in cellular biology research (because their DNA can be easily manipulated, they can double every 90 minutes and they even provide insight into human biological processes). It can be dehydrated down to 5 percent moisture content, be kept on a shelf and sealed with an inert gas and will last a year. Yeast is simply an amazing organism.

time. This creates the perfect environment for the production of alcohol greatly speeding up and concentrating biological activity. Lagers are cool fermented at 50 degrees Fahrenheit (and then stored for 30 days or more at close to the freezing point) while warm fermented beer is performed at temperatures of 64 degrees Fahrenheit

Wine production works the same way, except that yeast uses grape juice as the feedstock. While yeast is already present on grape skins, specific and pure yeast culture is added by the winery so that the process can be better controlled. Producing wine with minimum negative attributes means adding vitamins and minerals that support the yeast's metabolism.[38]

Wine has the degree of its fermentation tightly controlled. Dry wines allow more or full fermentation of the sugars, while sweet wines stop the fermentation earlier—leaving more

[38] Wikipedia

sugars.[39] Amazingly, wine was fermented for thousands of years without wine makers understanding the mechanism that produced wine.[40]

The previous two types of alcoholic beverages are fermented, but **do not** have to be distilled in order to increase their alcohol concentration. Beer can be distilled in order to remove solids such as spent grain and yeast.[41] On the other hand, beer can also be filtered to remove solids.

Beer and wine range from 5 to 12 percent alcohol. Fermentation completely stops once alcohol levels reach roughly 20 percent, which means getting a higher alcohol concentration requires distillation.[42] Distilled alcohols use the fermentation of grain, fruit or vegetables. A spirit is a distilled beverage that contains at least 20 percent alcohol by volume.

Distillation is the separation of the input into categories, or what is known in the industry as fractions, through the application of heat. A fraction is a grouping of compounds that boil at the same temperature. It is a physical separation process and not a chemical reaction and is applied to water, air, petroleum and alcohol.

Simple distillation is when the vapor is channeled into a condenser—therefore the distillate is not pure simple distillation but is useful for removing impurities—for instance simple distillation is used to distill water.

Fractional distillation was developed in the thirteenth century. Fractional distillation is the most common form of distillation and allows different fractions to be withdrawn at different boiling points along what is referred to as a distillation column. It is based upon the placement of diversion points where various molecules of the desired type congregate as a result of vaporization and condensation.

---

[39] Wikipedia

[40] Winemaking may be the oldest process industry-manufacturing sub-category. The oldest existing winery and winepress is 6000 years old in Armenia, and the oldest bottle of wine is 8000 years old. No one knows when winemaking was first performed.

[41] Alcohol Distillation: Basic Principles, Equipment, Performance, Relationships and Safety

[42] When wine is distilled it becomes what is referred to as brandy.

http://www.youtube.com/watch?v=Z6OyNB8V7Hc

Search Term: Fractional Distillation

The first book on distillation was published in Germany in 1512 (just 62 years after the development of the Gutenberg printing press—highlighting the importance of this topic).

Distillation can produce one product.

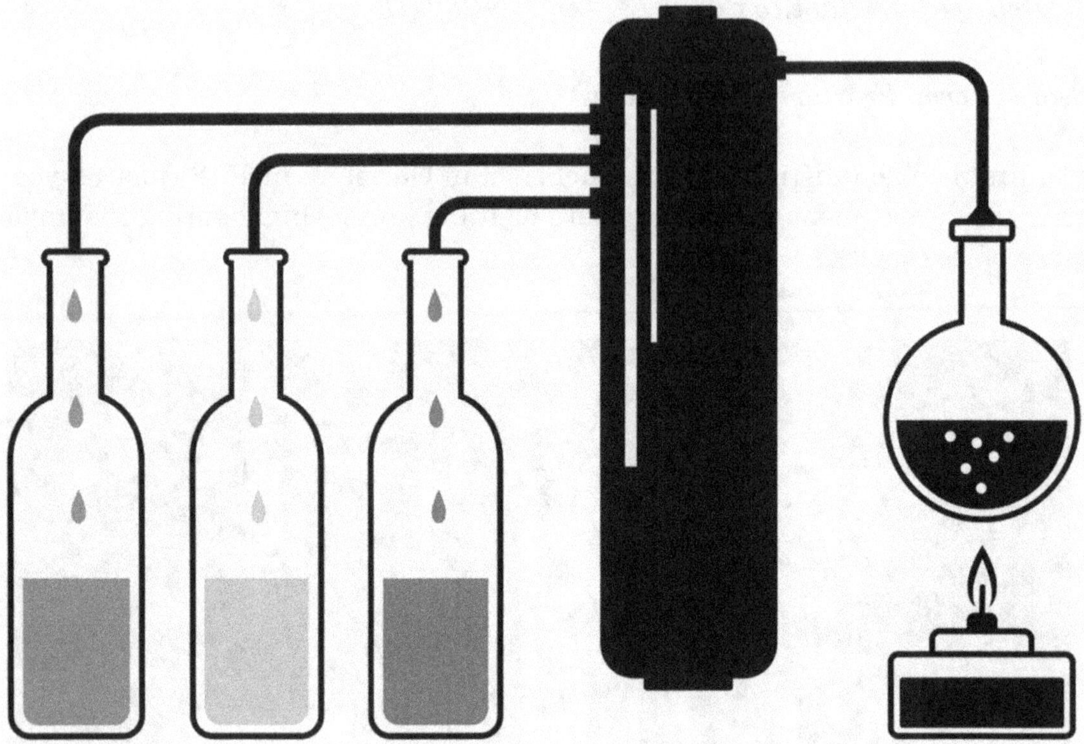

Distillation also can be used to derive more than one output product. This separates out each fraction by boiling point.

Dairy Farming

Cows are frequently milked in batches—with the batch being the categorization of cows of the herd by milking speed. Group A may be give up their milk in 5 minutes, while Group C may take 15 minutes to give up their milk. A herd may be milked between one and three times per day. This is one of the simpler process manufacturing problems to solve, although there can be quite a few cows, with many dairies having between 15,000 and 20,000 cows. In this environment, staging areas—called loafing barns—are setup prior to the milking parlor.[43] Dairy farming is much more focused on animal husbandry and waste management than

[43] Wikipedia

in planning milking/production activities. As such dairy farming is not much of a consumer of process manufacturing software—however, it is another example of batch processing.

http://www.youtube.com/watch?v=cPLTUnOOAN0

Search Term: Cow Milking

Paint

Paint production is performed in batch. This is because paint simply has too many variants to produce in a continuous manner. In addition to all of the paint colors, there are also many different paint types (such as gloss, matte, indoor, outdoor, etc.), pigments (for the paint's color), resins (to allow it to dry once it has been applied), solvents (to allow the various input products to mix properly), and various additives that make up the raw materials. In this example of batch production, the appropriate quantities of the paint inputs must be added, the mixing must be performed, the paint must be packaged, and then the paint vat must be cleaned out before a new combination of input materials can be added.

http://www.youtube.com/watch?v=hX_qTE5xlb0

Search Term: How Paint is Made

Glass

The first part of the process to manufacture glass is performed, appropriately enough, in something called a batch house. The batch house "*measures, assembles, mixes, and delivers the glass raw material recipe via an array of chutes, conveyers and scales to the furnace.*"[44] The furnace heats the glass raw material until it is plastic. A portion of the plastic glass input material is cut into a batch or bog. This material can then be blown (the process used to create glass items like bottles) or pressed in order to produce windows, covers for LCD screens, etc. This is not the end of the heat treatment. In order to promote even cooling, which will prevent the glass from cracking, the formed glass is placed into an annealing oven where it is repeatedly heated and then allowed to cool gradually.

[44] Wikipedia

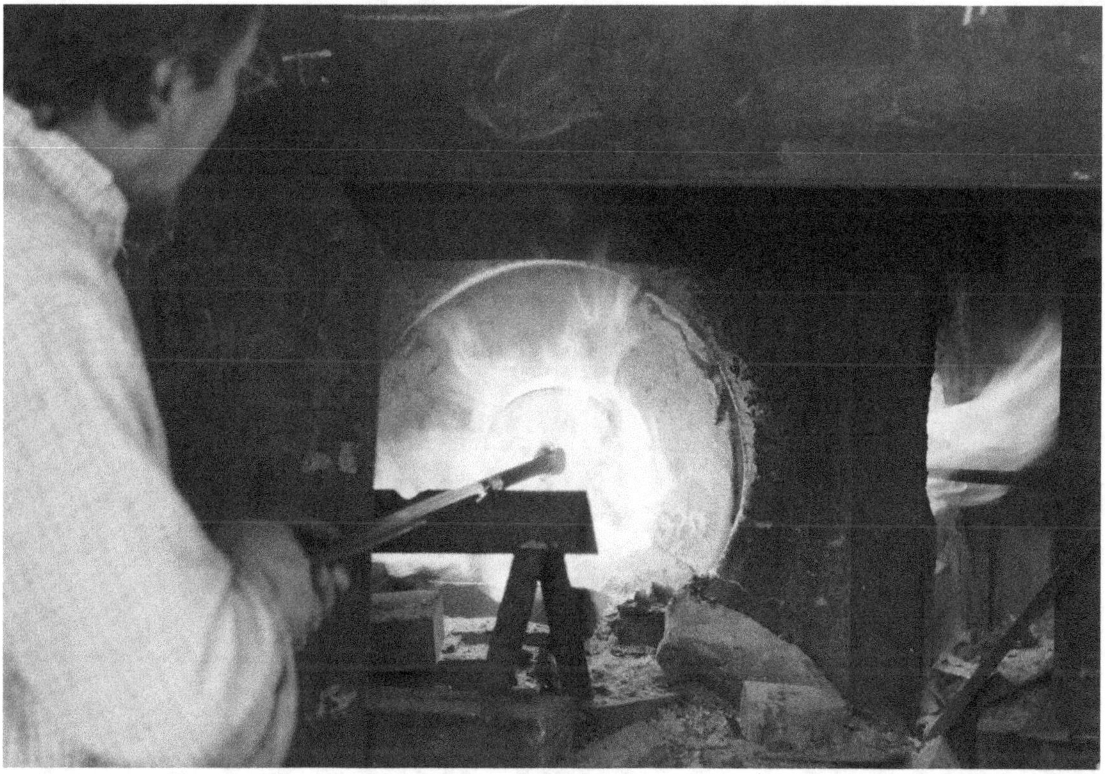

This is a smaller version of a glass furnace—the way glass used to be made.

After the "slug" is removed from the glass furnace it can be blown in order to make bottles, or laid flat in order to make sheets.

An implementing company that has either a discrete or repetitive manufacturing environment, will typically choose among the same software applications.

Specialty Chemicals

Specialty chemicals differ from commodity chemicals in that they are sold for their performance in application **rather than for their chemical composition.**[45] They are also different in that, as per their name, they have a specific use—often just one or two uses. Meanwhile commodity chemicals can have dozens of uses.[46] While a commodity chemical plant will normally make just a few chemicals, a specialty chemical plant may produce more than 100 different chemicals per year—and in fact specialty chemical manufacturers tend to be much smaller than companies that make commodity chemicals. To get an idea of the nature of specialty chemicals and how they have such detailed specifications, I recommend the Kodak Specialty Chemicals web page, which publishes their specifications.

> http://www.kodak.com/ek/US/en/Industrial_Materials_Group_LHN/Specialty_Chemicals/Products.htm

Specialty chemicals use commodity chemicals as their feedstock. Most specialty chemicals are organic chemicals, which means their molecules contain carbon. Specialty chemicals can have similar genealogy requirements such as pharmaceuticals. They sell for typically more than a dollar a pound.[47] This includes industrial gases, adhesives, sealants, coatings and catalysts. Commodity chemicals represent around one quarter of all dollar output globally of the chemical industry (but, of course, far more by volume).[48]

---

[45] Global Specialty Chemicals
[46] What is Specialty Chemical Manufacturing?
[47] Wikipedia
[48] Sectors of Chemical Industry

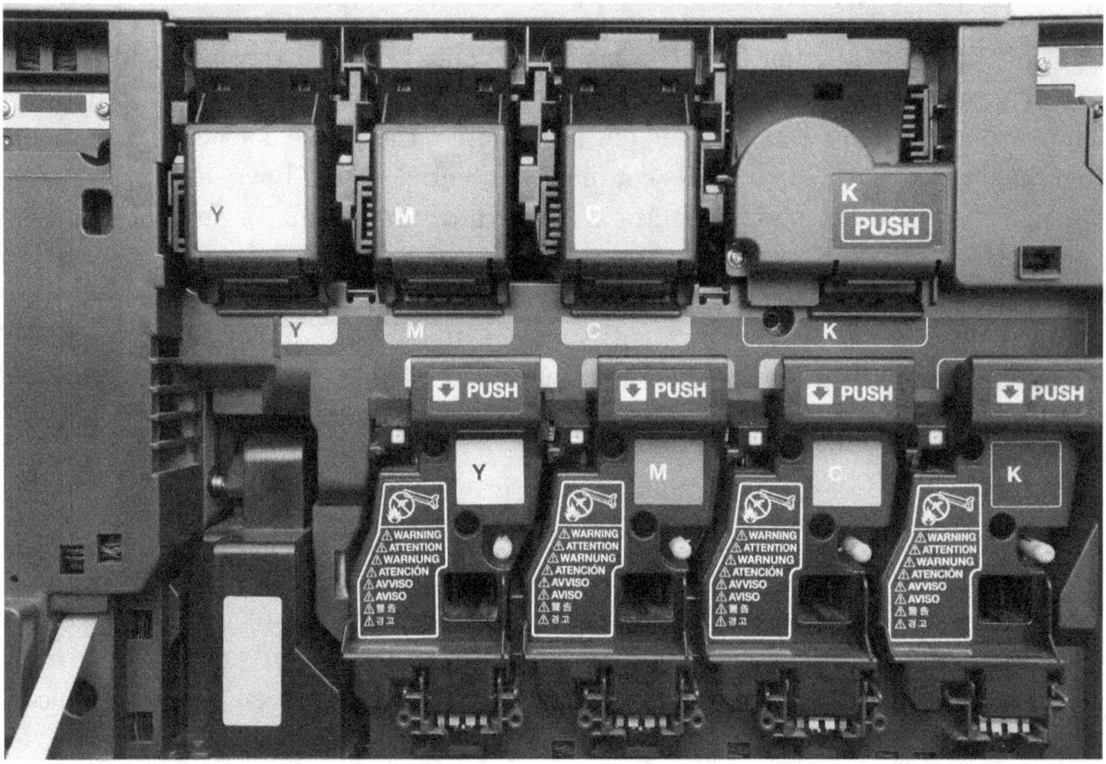

While it may seem simple, the ink in an inkjet printer is actually quite a bit more involved as the following quotation attests. "The process of inkjet printing is very complicated, and the ink used must meet certain chemical and physicochemical requirements including those related to storage stability; jetting performance; color management; wetting; and adhesion on substrates."[49] *And this ink is in fact a specialty chemical.*[50]

Pharmaceuticals

Pharmaceuticals are produced in batch because of the necessity to track each batch due to regulatory and safety issues. This is such a focus of pharmaceutical manufacturing that it is referred to as genealogy. Pharmaceuticals are so high profit, and the quantities of pharmaceutics that are manufactured is so low in terms of volume, while the regulatory and legal constraints on pharmaceuticals are so high that it is one of the few manufacturing areas where productivity is

[49] The Chemistry of Inkjet Inks

[50] Inkjets are actually used for a wide variety of industrial applications. They can be used to apply coatings. They generally excel at applying a specific quantity of material to a surface.

just not much of an issue. Applications that are directed toward pharmaceutical manufacturing provide full electronic batch records or EBRs, which enables analysis of deviations, quality and analysis. Pharmaceutical production planning and execution is specialized—and the actual production optimization, versus the tracking, archiving and regulatory compliance is small, and this is why pharmaceutical companies tend to look for solutions that are customized for their needs. In fact, a major issue with these types of systems is the degree to which they can be made paperless—so that the administration burden of producing pharmaceuticals can be reduced.

Some people propose that pharmaceuticals will be able to move increasingly to continuous production and away from batch production. On the other hand pharmaceuticals are such a low volume area of manufacturing, it's hard to see the point. For example, Exxon has one refinery that processes 650,000 barrels of crude oil a day. That is more material than all the pharmaceutical companies in the world put together would run in a year. Because of the quality and traceability constraints in drug manufacturing, and the relatively small quantities involved, there is no overriding reason to convert the manufacturing process to continuous. In pharmaceuticals, the paperwork is more important than the efficiency of the manufacturing process.[51]

Moving Between Batch and Continuous Manufacturing

Some methods of production that at one point were manufactured in batch were converted to continuous through improvements to process control. There are multiple examples of this in both chemical and petroleum refining.

> *"For 30 years after the discovery of crude oil, refining consisted of separating these various products by batch processing, tediously handling one tank of crude at a time. Starting around 1900, refiners strung these tanks in series and used a so-called continuous batch process, still a capital and energy intensive way to do separation. Fractional distillation, using the trayed columns now used worldwide did not come into widespread use...for another two decades."*
> — Petroleum Refining in NonTechnical Language

Continuous production has often-quantifiable productivity improvements as is demonstrated by the following quotation from this same book, which is referring to the change in oil refining as it moved from batch to continuous production.

> *"The efficiency of separating crude oil into its constituents increased by 25%."* — Petroleum Refining in NonTechnical Language

What is interesting is that while in discrete manufacturing the trends are to attempt to create "flexible" manufacturing environments that result in shorter production runs and more changeovers, in process industry manufacturing the emphasis is, whenever possible, to move away from batch production to continuous production. Continuous manufacturing normally has the highest degree of equipment investment. For instance, *"process units can cost from $50 million to $1 billion.*

[51] Dave Biddle, Senior Application Consultant, PlanetTogether

A paper machine is on the order of $250 million to $1 billion."[52] Although this is not always the case; some chemical production that was running at high production rates were converted back to batch processing in order to meet the market demand for product variability. A major impetus to transition from continuous to batch manufacturing is product proliferation, as the following quotation explains.

> *"Today, process manufacturers have a proliferation of products. These products are produced using the same equipment or using existing processes downstream but at the same site. Because of the number of products now flowing through the same cost center, using financially-based cost systems to develop accurate, individual product costs is difficult, sometimes impossible."* — Process Industry ERP Requirements

The efficiencies that come from continuous manufacturing are only useful if the demand consistently exists to consume the output. Continuous operations are easier to model than batch processes. This is explained in the quotation below from PlanetTogether.

> *"In a multi-stage production, it's often the case that material flows out of one piece of equipment directly into another with no storage in between. For example, a hot dog manufacturer may grind beef in the grinder while the ground beef is fed directly into a stuff line. If each grinder has a one-to-one fixed connection to one stuff line then the process can simply be modeled as a single operation. However if each grinder makes products that can feed multiple varying stuff lines then this presents a scheduling challenge. As material is made by one operation, it is creating intermediate material at a particular rate. This material is then consumed at a particular (perhaps varying) rate. In our example, a grinder may be making beef that is consumed by multiple stuff lines simultaneously (which perhaps start at different times and run at different rates). This is handled by PlanetTogether by automatically examining the cumulative production and consumption rates of the material over*

[52] Dave Biddle, Senior Application Consultant, PlanetTogether

> *time and ensuring that the resulting schedule is feasible that enough beef is being ground to feed the pack lines as scheduled."* — Using PlanetTogether for Effective Scheduling in the Process Industry

Regarding the above quotation: you can have a continuous operation (like an oil refinery), but the down-stream disposition (use) of the materials can be routed to many different process units to make many different products as required.[53]

Conclusion

Process industry manufacturing is divided into batch and continuous production, and these categories correlate to discrete and repetitive in terms of their constancy of production. The differentiation is whether the finished good can be disassembled after production into its component parts.

Continuous and repetitive manufacturing is the most efficient method of manufacturing. This is in both manufacturing environments, they have very little downtime, and the continuous nature of these manufacturing processes is able to leverage large capital investments into highly automated machinery. Although sometimes misunderstood as such, continuous manufacturing does not mean that the exact same product is produced. A variety of manufacturing adjustments can be made without performing a changeover that changes the type of material produced. The concept of requiring a changeover in order to produce a new product is really a holdover from discrete and batch manufacturing. On the other hand, in continuous manufacturing, there can be more than one way to make the same finished product.

In process industries, "batch production" is where the production process is stopped and restarted in order to produce another batch of the same product or of a different product. Some methods of production that at one point were produced in batch were converted to continuous production through improvements to process control.

[53] Dave Biddle, Senior Application Consultant, PlanetTogether

CHAPTER 3

The Discrete Manufacturing Focus of Most Manufacturing Software

Manufacturing software tends to be targeted toward discrete manufacturing, often leaving process industry manufacturing without software solutions that are designed for its specific requirements. To meet this gap in functionality, software vendors will develop a solution for discrete manufacturing and then simply sell it into the other manufacturing environments for which it is a poor match. Many software vendors are filled with salespeople who are under pressure to sell no matter what, and have little to no concern about what happens to the project during implementation. Here's a typical scenario, it is repeated so frequently that it has steps as in an algorithm.

1. The software vendor proposes to executive decision makers that their discrete solution can meet all of their needs. The executive decision makers—often misadvised by consulting companies who stand to make money off of the implementation following a particular brand name software vendor, and the IT department that prefers familiar software vendors over software that meets the business requirements, purchase the inappropriate software.

2. After the software fails, the company will bring in consultant after consultant to help them fix the problem.
3. If the consultant is dishonest (not hard to find by the way), he or she will try to tinker with the system (while billing of course), and may make some small improvements, but the application will still fail to meet the customer's requirements.
4. If the consultant is upfront, he or she tells the company that they purchased the wrong software and that they need to look at their implementation as a "sunk cost" (and then confronts their consulting partner about why they are providing the company with such poor-quality advice).
5. The company will replace that consultant with someone who makes modest adjustments, but most importantly supports the view that the purchase decision was a good one.
6. When all else fails, blame the users for not "understanding" or "learning" the inappropriate software. The executive decision makers, the software salesmen, the consulting company, IT—they mostly come out scot-free for either supporting or making the poor decision.
7. Repeat

The process of attempting to recover software that never should have been implemented usually goes on for years, and is very damaging to companies. I recall presenting a detailed analysis of why an application that the client had implemented could never meet their requirements. The client could not refute anything I presented, but afterward turned to the consultants from a major consulting company—who had recommended the application—and asked them what they thought. The major consulting company—who made their money by recommending the solution, as it is the only application for which they are staffed—told the client "they would look into it."

Improving an application with a design that cannot account for process industry manufacturing requirements is difficult, and many large vendors have little incentive to change because all the major consulting companies still recommend their software no matter what the major/brand name software vendor offers. Most of the software vendors are quite familiar with the limitations of their software in process industry environments, and do not allocate resources to improve the software.

This has gone on for years, and many companies have not been able to figure this out.[54] I can't explain this, because it is entirely obvious. I have seen applications that were never designed to meet process industry requirements continue to be sold into process industry accounts. The previous failures of the application with similar clients never seems to affect the ability to sell that product to new process industry clients. This is particularly true of software vendors that have a major brand in the enterprise software market.

One problem that leads to all of this is that there is no entity, at the time of this publication, that rates manufacturing software in sufficient detail to help companies make informed decisions between alternatives. If we look at Gartner (the well-known IT analyst firm that many companies use to assist in their software selection), their ratings or what they call "Magic Quadrants" **are limited to supply chain planning suites**, which include applications like demand planning, supply planning, etc. Companies cannot turn to consulting companies for advice, because they recommend the software in which their consultants already have experience and with which they have established relationships.

Many years of analysis demonstrate conclusively that consulting companies do not have an interest in, or exposure to, the best applications that could add value for their clients. The standard consulting policy regarding software recommendation

[54] Interestingly, the focus of software vendors on discrete manufacturing extends to ERP software, as the following quotation explains.

> *"In the race to fix these problems, companies often lose sight of the fact that ERP packages are nothing more than generic representations of the ways a typical company does business. While most packages are exhaustively comprehensive, each industry has its quirks that make it unique. Most ERP systems were designed to be used by discreet manufacturing companies (who make physical things that can be counted), which immediately left all the process manufacturers (oil, chemical and utility companies that measure their products by flow rather than individual units) out in the cold. Each of these industries has struggled with the different ERP vendors to modify core ERP programs to their needs."*—The ABCs of ERP

is set at the highest levels in those consulting companies, and they are based upon revenue maximization.[55]

This problem in adapting discrete manufacturing software to process industries is explained in the quotation below:

> *"Supply chain professionals face a frustrating task when they try to use the principles articulated in the literature and the software available on the market. At a conceptual level, there are many similarities between the discrete and process environment. But agreement at the conceptual level does not necessarily translate to successful implementation. When software systems and practices that do not account for fundamental process characteristics are mandated, it usually leads to the development of parallel systems because the folks on the ground have to deal with practicalities. In many companies the planning tasks and supply chain management continues to be done in an informal way with spreadsheets and desktop tools, with the results being fed to the officially mandated supply chain system."* — Supply Chain Planning in the Process Industry

In many cases, the executives at companies cannot differentiate between supply and production planning systems that can meet the requirements of process and repetitive manufacturing. Instead they make their selection decisions based upon larger, more "strategic" considerations (such as which vendor has the best-known brand, or attempting to buy applications from the same vendor). As a result, the

---

[55] In Australia, when I worked for i2 Technologies as a Project Manager, I objected to the placement of unqualified consultants from PwC onto the project. PwC attempted to get me to sign off on a budget for more consulting that the client did not need. When I rebuffed PwC, they attempted to have me removed from the project, and stated that not only would they retaliate against i2 Technologies in Australia by not recommending i2 Technologies' software to their clients, but that they would retaliate across all of Asia in a similar manner. This is clearly indicative that the PwC Partner and Associate Partner were far more interested in maximizing their consulting revenues versus recommending software that was good for their clients.

failure rate of supply and production planning implementations at process and repetitive manufacturing clients tends to be high. So the software does get implemented, but it just never gets used—or is partially used.

Common Functionality for the Process Industry

So what common and important capabilities are typical of process industry manufacturing? You can gain insight to the requirements for process manufacturing planning by observing a company called TGI, which declared the following capabilities within its Enterprise 21 software application.

- *"Supports infinite-level formulas with yielding at the top level or ingredient level and rank-ordered ingredient substitutes*
- *Products can have global formulas or unique facility-specific formulas*
- *Supports scalable batches*
- *Supports recipes through the integration of formulas and associated processing instructions and notes for use during batch process production*
- *System maintained version and revision control with a fully integrated engineering change order process*
- *Formulas support nested formulas and intermediates and can have infinite notes and instructions*
- *Supports online review of formulas via single-level and multi-level explosion*
- *Multi-level where-used functionality enables rapid access and mass change to all finished goods using specific ingredients, intermediates, and nested formulas*
- *Formulas and recipes are easily duplicated and updated for the same product or a new product*
- *Formulas are pulled into work orders or batches for tracking and recording of actual ingredients consumed or backflushed at defined standards*
- *Supports co-products and by-products"*

Why Companies End Up With Applications That Don't Meet Their Process Requirements

I have actually worked at several clients where the company's executives did not understand their own manufacturing and supply planning processes. Here's an example of what happened at one such client.

A software vendor that had weak production planning and scheduling capability for process manufacturing told the company's executives that their applications contained all of the industry's "best practices." Hearing this, the client never checked the actual capabilities and functionality of the applications. During the implementation, the executives (who again did not understand their business processes) told their own business to "get with the program" and change their antiquated ways (which were in fact not antiquated; they were processes that were designed for the specific requirements of their factories, rather than the generic functionality that was available for the software). The executives insisted that the company change its processes to meet the "best practices" (i.e., lack of specific functionality for their actual process needs) so that the software could be implemented. This is a case where change and transformation terminology were used as excuses to force changes that, if successful, would have undermined the planning effectiveness of the business.

Because this scenario has repeated itself in so many companies, a significant percentage of companies are using planning systems that were ***never designed for their environment***. The end result is that those who work in process industry manufacturing believe that off-the-shelf applications cannot meet their requirements, when in fact their companies have simply been making poor software selection decisions. This goes beyond external planning systems to ERP, as the following quotation attests.

> *"For IT chiefs like Pam Haney, IT director at Irvine Scientific, a $28 million life sciences company, issues of competitive advantage are overshadowed by ever-present customization needs with her ERP system. 'What we're faced with is having to deal with ERP packages that are not designed for process manufacturers.'"* — Why ERP Systems are More Important Than Ever, CIO

Poor Software Selection as a Contributing Factor to the High Failure Rate of Enterprise Software Implementations

Poor software selection is one of the reasons why enterprise software implementations have such a high failure rate. This issue is covered in the SCM Focus Press book *Enterprise Software Selection: How to Pinpoint the Perfect Software Solution using Multiple Information Sources,* but it's important to explain here in this chapter a few things about how software selections can go sideways, mainly because the market for manufacturing software is a perfect example of the continual problem that companies have in selecting the appropriate software for their needs.

1. *Buyers are Often More Interested in Factors Outside of Implement-ability*: The enterprise software market is not an efficient market. Many factors drive buyer revenues between software vendors, and the implement-ability of software is only one of the factors.

2. *Buyers Listened to Advice from Advisors Who Had Conflicts of Interest*: As an example, many large vendors have partnerships with large consulting companies. These large consulting companies continue to recommend inappropriate software for no other reason than it maximizes their revenues. Smaller software vendors will never enjoy this relationship with large consulting companies regardless of how good or implementable their software is. Once a consulting company grows beyond a certain size, it becomes all about sales numbers—and the consulting company is more interested in maximizing its business, not the business of its clients.

3. *Information on Software Successes is Skewed*: Software failures are hidden, therefore the natural feedback loop—which would provide information about software that implements better than other software—is to a great degree broken.

4. *Buyers Want Large Brand Name Vendors*: IT in particular wants as little vendor diversity as possible because less diversity reduces their workload—or at least they think it does. IT is not compensated or motivated to meet actual business requirements. Even within the company purchasing the software, some IT employees have more allegiance to their preferred software vendors than they have to the company they work for. However, in many cases large brand name vendors do not have the desired functionality. Process industry

manufacturing planning is a perfect example of this situation. This is a highly specialized segment of manufacturing planning that requires deep domain expertise, and its needs are better met by smaller vendors.

Custom-coded Solutions

Custom-coded solutions can be created using general optimization solvers such as CPLEX or the optimizer to MATLAB. Because these applications are not pre-configured, they are highly customizable. For instance, I recently read about a custom production scheduling system that was designed for a smelting operation. However, the downside is that custom solutions are expensive and take a long time to develop, require specific mathematical and mathematical development and skills to be successful, and can have significant ongoing maintenance costs. Finally, custom solutions must be 100 percent maintained by the in-house support staff.

Background on Custom Solvers

There is an important distinction between optimizers in enterprise software versus optimizer engines/solvers in modeling languages. The following quote is instructive:

> *"The reason for preferring C and C++ may also lie in the fact that in software companies such as SAP or i2 Technologies, far more people are found with a computer science background than a mathematical programming background. In the latter community, modeling languages are more frequently used supporting the concept of strictly separating the data, model and solution methods."* — Real Optimization with SAP APO

> *"However, if SAP APO would become open in the sense that the user can access the variables and constraints, a modeling language would definitely be the better choice. A modeling language would also make it easier to modify and extend the predefined model."* — Real Optimization with SAP APO

Open modeling languages/solvers include the applications mentioned above: CPLEX and MATLAB. In these applications the actual mathematics are visible to the user and can be adjusted. With enterprise software, the mathematics are proprietary and the implementing company can only understand the optimizer by referring to the documentation that the vendor makes available and by observing the results of optimization runs. This is a fundamental difference between enterprise software optimizers and modeling languages. Other differences are also explained:

> *"When it comes to optimization, standard business software products such as SAP APO do not offer LP and MILP modeling per se on the equation level, but provide a standard MILP formulation suitable for a wide variety of real-world problems that is configured by manipulating data objects in the business application. Hardly anywhere are terms such as variable or constraint found. The modeling parameters are set by defining supply chain related objects such as products location, transportation lanes, lot sizes, resource capacities, etc."* — Real Optimization with SAP APO

Therefore the modeling is stated in very business-oriented terms, with the actual mathematics hidden and the terminology of modeling languages like MATLAB essentially unused.

The Difference Between Enterprise Optimizers and Modeling Language/Solvers

Enterprise optimizers and modeling languages/solvers are really designed to support different processes. Enterprise optimizers are designed for canned implementations where the company looks to implement a production optimizer to run the business with limited visibility into what the optimizer is doing. Modeling languages/solvers are designed to offer an unlimited amount of customization and often have analytical tools (MATLAB offers graphical capabilities so the optimizer can be viewed in terms of its performance as well as having its mathematical output analyzed) and a big part of this is sensitivity analysis. Additionally, enterprise optimizers are typically not a good choice for problems that are very specialized, and a custom written optimizer in a modeling language/solver is highly preferable over using a preconfigured enterprise optimizer.

In general, my view is that enterprise optimizers tend to be over emphasized at the expense of modeling languages/solvers. Enterprise optimizers can be used for strategic analysis (and there are significant differences between various enterprise optimizers in this regard), but performing simulations with enterprise optimizers is more time-consuming and expensive than with modeling languages/solvers. However, it is not necessary to choose between the two. A modeling language/ solver can be used in conjunction with an enterprise solver in order to gain the flexibility and speed of adjustment that enterprise optimizers tend to lack.

The Production Planning and Scheduling Sweet Spot

Let's take the example of one of the five software categories that are covered in this book; production planning and scheduling software. I have found that the current thinking on production planning and scheduling is that it requires sophisticated software. There are a lot of examples where this is true. From what I have seen, there seems to be a "sweet spot" where most "standard" advanced production scheduling software seems to apply, especially when the scheduling problem is somewhat complex but not extremely complex. Secondly, different scheduling software will work or not work based upon the complexity of the problem and the type of scheduling that is performed. This relationship is shown in the following graphic.

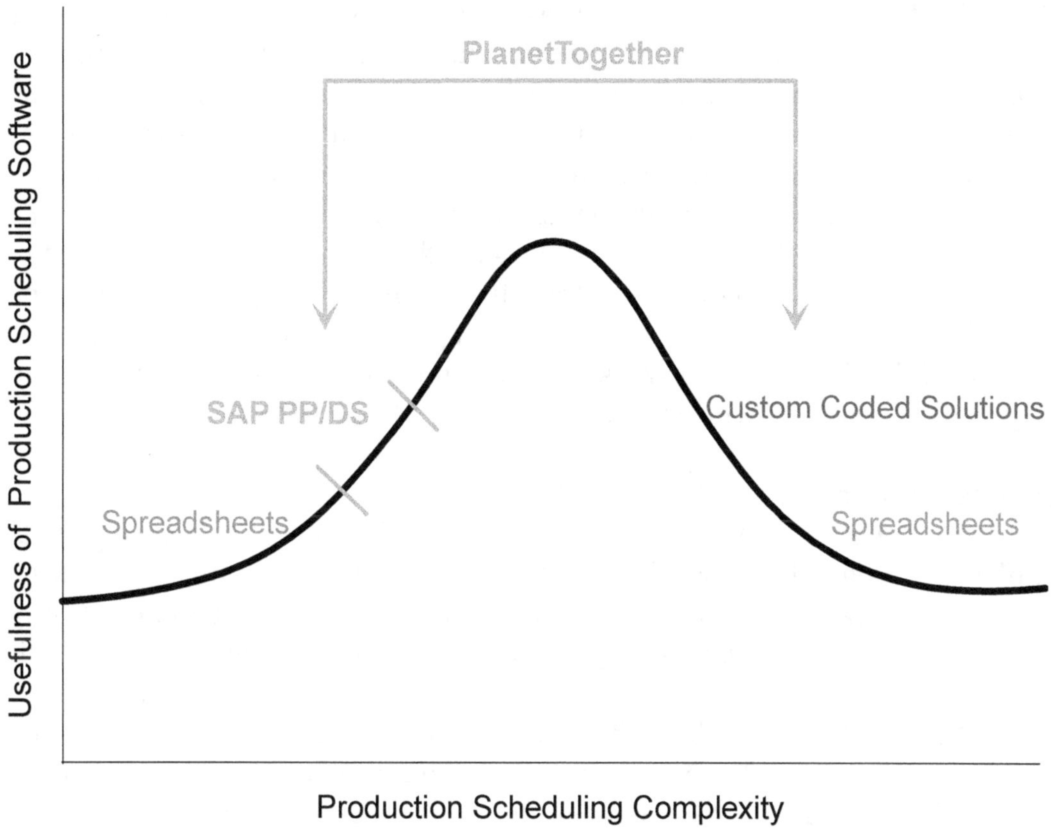

This graphic demonstrates the "sweet spot" in the relationship between packaged production planning and scheduling software. As noted above, the sweet spot is located between when a scheduling problem is very simple and very complex. This relationship also has an overlay with different applications. For instance, spreadsheets tend to be better for very simple or very complex problems. However, within the scheduling software sweet spot, a weak production scheduling application like SAP PP/DS (which incidentally is representative of most production planning and scheduling applications in the market) has a narrow band where it can be implemented successfully.

PlanetTogether and AspenTech, on the other hand, are best-of-breed applications that are far more appropriate for process industry manufacturing planning. They have a wide band of production scheduling problems over which they can be implemented; however the band is also not unlimited. Because the process industry has such a wide variety of manufacturing requirements, no packaged or off-the-shelf application can meet the requirements in all circumstances. Further on in this book, as we get into detail regarding the specific process industry requirements, it will become evident why this is the case. From far enough away, all process plants look the same. But the truth is, when you look closely enough, you'll find that they differ in important ways—even a plant that's owned by the same company and replicated down the street.[56]

The Lack of ERP Systems for Process Industries

It is not only production planning systems that often have been developed from requirements outside of process industry manufacturing and sold into process industry manufacturing companies. ERP vendors have been adamant that their software can be used for any industry. SAP has many different industry solutions. However, when one looks closer at these proposed capabilities there is often much less to the actual functionality. I have worked on several repetitive manufacturing SAP implementations. Repetitive manufacturing is where the production is continuous; the equipment investments are high as is the rate of production. However, on these projects it finally came out, late in the project of course, that the repetitive manufacturing functionality is **not** really worth turning on. Talking up the "repetitive manufacturing solution" enabled SAP to get the sale, but wasted an enormous amount of time on this project—and on many other projects, because of the constant project discussions as to whether repetitive manufacturing would be used. As a very experienced SAP consultant, from SAP itself once stated to me "*there is really more sizzle than steak to SAP repetitive manufacturing.*"

This issue of there not being much there is actually repeated throughout the SAP industry solutions—which are completely divergent from the SAP website and other SAP marketing material that make it appear as if SAP applications have all of these customized templates and customized functionality to meet every

[56] Dave Biddle, Senior Application Consultant, PlanetTogether

possible industry's needs—including both the overall process industry, as well as sub-categories of the process industry.

There is a general misunderstanding on this topic among implementing companies. Just because a software vendor has a number of customers in a particular industry, does not mean that this software serves that market or provides the functionality that is necessary—this is certainly the case when so much of the ERP system is customized. Many executives simply buy software because other companies buy it, for career reasons or because the marketing and the salesmanship dazzle them. In fact, many companies do not analyze the actual investments they make very carefully, preferring to simply pattern on what competitors are doing as the following quotation attests.

> *"Interviews with the managers confirmed that if a successful strategic IT (e.g., laptops to salespeople) was implemented by one or two firms the other competitors soon followed the lead. Thus although the IT intensity of the industry increased, no net performance effect was observed."* — The Relationship Between Investment in Information Technology and Firm Performance: A Study of the Valve Manufacturing Sector. Information Systems Research

These SAP industry solutions are ***really more marketing constructs than actual usable functionality*** and because the marketing is so good, it took me years of working on SAP projects to figure this out. I have not worked with every industry solution, but I have worked with many of them, and none of them have worked as advertised, and most of the time the ERP system is implemented without using the functionality of the "industry solution." SAP industry solutions are designed to get executives in purchasing companies to think that SAP can meet the unique requirements within **their** industry, but in truth, most ERP systems offer generic functionality—as if all companies are the same—and if a company wants or needs something specific—they need to customize the ERP system. Another common division between manufacturing companies, who are the most common implementers of ERP systems, is the division between discrete and process industry manufacturers.

In SAP ERP, there is a functionality, which can be enabled, called batch functionality, which is used for batch process manufacturing. It is a very popular requirement for all types of batch manufacturing. This is explained in the following quotation from ProcessPro, a process industry ERP software vendor.

> *"While any company that utilizes process manufacturing could benefit from ERP lot traceability, companies regulated by the Food and Drug Administration (FDA) are required to be prepared for a recall at any given time. With a lot traceability feature, companies are able to easily isolate a food safety issue. In doing so, allowing them to forecast the extent a product has been affected, including what items have been contaminated and where they originated."* — Solving Six Common Challenges with ERP

Batch manufacturing can be enabled in SAP ERP as is described in the following link.

> http://help.sap.com/saphelp_SCM700_ehp01/helpdata/en/48/aae32a8740356ce10000000a421937/frameset.htm

However, it is well known as a major headache to enable. This is because batch management was added on as an afterthought to what is foundationally not only not a batch management ERP system, but also not even a process industry ERP system. SAP ERP was designed to work with discrete manufacturing—which simply has different requirements.

Businesses require solutions that have been designed from the ground up to meet their requirements, not software that attempts to and successfully appeals to all industries with slick marketing literature and the barest level of functionality so that an ERP vendor may claim that it can address the specific requirements. Right now, there are thousands of companies that have implemented unplanned customizations into their ERP system because the ERP vendor told these companies that they could cover their industry specific requirements when in fact they could not. With so much money flowing into Tier 1 ERP vendors that have provided generic solutions and unfulfilled industry specific functionality promises, it has prevented money from flowing to solutions that could have been developed to offer more companies software that meets their requirements in more than marketing literature. This is another example of a prediction make by many ERP vendors that helped them get sales, but has been not only bad for their "customers," but bad for the enterprise software market overall.

Conclusion

There are a number of manufacturing environments, and each must be managed with supply and production planning approaches that fit those specific environments. The first step in choosing the right software for the various process industry environments is to understand the differences, and then use this information to develop strategies that match the needs of each environment. Most software is designed around discrete manufacturing, and most books and consultants also speak from the perspective of discrete manufacturing. But techniques taken from discrete manufacturing often don't make any sense to apply to process manufacturing. Managing each environment properly means going beyond simple solutions.

Manufacturing accounting, manufacturing planning, manufacturing execution scheduling, and process control encompasses a large number of problems, and within this continuum there are solutions that are better fits than others. These areas are not a "one-size-fits-all," regardless of the intent and financial backing of the major vendors, and all the major consulting companies who present the area in this oversimplified way. In order to make an appropriate software selection, it is necessary first to analyze the specific production and scheduling problem and the company's specific detailed requirements, and then to compare them to the alternatives that are the best fit for them. This type of due diligence is very rarely performed, and the repeated issues that companies have in making good software selection decisions and in interpreting biased advice that they receive was a major motivator for the SCM Focus Press book, *Enterprise Software Selection: How to Pinpoint the Perfect Software Solution using Multiple Information Sources.* Because so few consulting companies focus on process industry manufacturing or treat it differently from discrete, it is difficult for companies to get quality outside support on software selection.

Manufacturing software, be it ERP, planning, MES or process control tend to be targeted toward discrete manufacturing, often leaving process industry manufacturing companies without software that is designed for their specific requirements. In many cases, the executives at companies cannot differentiate between supply and production planning systems that can meet the requirements of discrete and process manufacturing. Instead they make their selection decisions based upon larger, more "strategic" considerations (such as which vendor has the best-known brand, or attempting to buy applications from the same vendor). Because this scenario has repeated itself in so many companies, a significant percentage of companies are using manufacturing systems that were never designed for their environment and that have superficial and incomplete functionality, designed more to sell into process industry accounts than to be successfully implemented.

Unfortunately it is not only production planning systems that often have been developed from requirements outside of process industry manufacturing and sold into process industry manufacturing companies. ERP vendors have been adamant that their software can be used for any industry. This has caused many implementation issues and disappointment in process industry companies that find themselves investing in the customization of ERP systems designed for discrete manufacturing, that were sold to them as process industry compliant when they never were.

CHAPTER 4

Co-Products and By-Products

Unlike discrete or repetitive manufacturing, process-manufacturing industries have associated output that is produced as a natural consequence of the production process to produce the intended finished good. Co-products are valuable outputs to the manufacturing process and they must be co-planned along with the primary output product.

By-products are unwanted outputs of the manufacturing process. Some can be sold at a low price, and others are actually a cost as they are simply a waste product and as they cannot be sold must be disposed of (there is some variability in the industry as completely unusable output is also called waste). Examples of by-products include animal parts, feathers, manure, poultry litter, whey, brewer's yeast, grape seed oil, molasses, orange oil, sawdust, mineral oil, salt, slag, among many others.[57] A constant by-product of natural gas and petroleum

[57] Interestingly, before the internal combustion engine was developed, anything but kerosene from oil refining was considered a by-product. Another common by-product was natural gas, which was previously burned off as it was considered valueless. It is now either processed as explained in Chapter 2: "Continuous Versus Batch Production," or re-injected into reservoirs. That is a story that is often repeated; as time passes the process industry becomes more creative, often through implementing more processing, about converting by-products to co-products.

refining is sulfur, as was discussed in Chapter 2: "Continuous Versus Batch Production." Because petroleum refining is performed on such a massive scale, sulfur is no longer mined. In fact, almost all-elemental sulfur is produced as a by-product from these two process industry sub-categories. While sulfur is a nuisance to petroleum refining (high sulfur crude oil is called "sour," while low sulfur crude oil is referred to as "sweet"), once separated from the crude, it is useful in fertilizer (mixed in equal quantities with phosphorus) and for the production of sulfuric acid—a primary component of many uses in the chemical industry[58], in matches, sulfur dioxide gas (for winemaking), insecticides (many organic insecticides contain nothing but sulfur), fungicides, gunpowder and for vulcanizing rubber.[59] This means modeling the co-products and by-products so that the optimal combination of these outputs, along with the intended output product can be produced, and then following through on the plan by controlling the production process, sometimes with process control software, sometimes without it, in order to match the planned production.

ERP Implications

The ERP system needs to be able to allocate costing appropriately. For instance, while both the primary product and the co-product receive costs, by-products do not have costs allocated to them because they are not a sold product. For instance, in SAP ERP the following applies to co-product and by-product costing.

In SAP ERP, co-products and by-products are set in the material master. Both co-products and by-products are represented by negative values.

1. *"The costs for co-products are calculated using the apportionment method.*
2. *The costs for by-products are calculated using the net realizable-value method.*
3. *The costs for fixed-price co-products and by-products are subtracted from the total costs.*

[58] Sulfuric acid is used both in chemical production in mineral processing, oil refining, wastewater processing for chemical synthesis. It is also a component in drain cleaners, various cleaning agents and lead-acid batteries. Sulfuric acid is used to remove oxidation from rolled sheet iron and steel (see Chapter 2: "Continuous Versus Batch Production"). Sulfuric acid is also an input product into aluminum sulfate, which is used in papermaking.

[59] Wikipedia

4. *After the costs for by-products and fixed-price co-products have been taken into account, the total costs of the production process are apportioned for all cost components to the co-products.*

5. *The system also creates an itemization and a costed multilevel BOM. When the costing result of a co-product is displayed, the other co-products are represented as individual lines with item category A. In such items of category A, the output quantity of the co-product has a negative quantity and negative value. The negative value is equal to the cost portion of the co-product that was calculated through the apportionment.*

6. *To cost a co-product using the recipe of a process material, you must create a costing view for the process material. For process materials (material type PROC), information is stored in the costing view of the material master for use when costing the co-products."*[60]

This is just one of the types of process industry requirements for ERP that is covered in Chapter 11: "ERP for Process Industry Manufacturing."

[60] SAP Help

CHAPTER 5

Recipe Management

The previous chapter explained the problem with confusing terminology between the various types of manufacturing. Terminology issues arise again in this chapter, which deals with the recipes, formulas and the bill of materials, or the BOM.

What Is a BOM?

A BOM is simply a list of input items that connect in a predefined way to an output item in certain proportions and that takes a certain amount of time to be converted. BOMs are used for discrete and repetitive manufacturing and not for process industry manufacturing. The BOM is one of the most frequently changed master data elements in the ERP system. However, it is beneficial to compare and contrast the BOM to recipes and formulas, as this is one of the main areas of differentiation between discrete, repetitive and process industry manufacturing. A BOM should never be originated or managed in any ERP system, planning system, MES or process control system (although the last example would not really be possible in any case) but instead should be managed in a system designed specifically for its needs, and then sent to the manufacturing systems.

For discrete manufacturing (e.g., automobiles, toys and tools), the BOM relationship is between multiple input items and a single output item. However, in process manufacturing, where the final item cannot be broken down and converted back to the original input products (i.e., cheese cannot be disassembled into its original components, but an automobile can), one input item can convert to multiple output items, or multiple input items can convert to multiple output items. All of these relationships can be easily modeled in a spreadsheet. A more extensive explanation is available in the SCM Focus Press book *Supply Planning with MRP, DRP and APS Software*. The example below is from this book and uses an actual system file format from the demand and supply planning vendor Demand Works.

Demand Works Smoothie

from_item	to_item	link_type	offset_days	factor
BEER	ALCOHOL	0	0	0.050
WINE	ALCOHOL	0	0	0.120
LIQUOR	ALCOHOL	0	0	0.400

This is a process industry BOM because it has a factor (however, Demand Works uses the same term or BOM to describe it as it uses the same table—but a "recipe" leverages functionality that the BOM does not as is described), which is less than one, meaning that there is a conversion between quantities and they are a percentage. The "to_item" is Alcohol, which is converting to the "from_item," which is Beer, Wine, and Liquor. More Alcohol is used in the production of Liquor than Beer, and this is only one of the ingredients to each of these output products. The "offset_days" is currently set to zero, but in reality, the manufacturing process takes time and should be higher than zero.

Understanding BOM Explosion

"Explosion" essentially refers to the multiplicative calculation performed by the planning system method, in this case MRP. In the example above, if 1,000 items of liquor are demanded, then 400 units of alcohol (1,000 x 0.4) are demanded, in addition to a similar multiplication of 1,000 for any other items that are also connected to the Liquor from_item in the BOM (not shown in the simplified BOM above). Explosion could just as well be called "finished good to input product relationship multiplication." The input products to the BOM are the dependent demand. Their demand is "dependent" upon the demand of the finished good.

BOMs, Recipes and Formulas

The software's table I showed on the previous page is called the BOM table. In it, I placed a process industry product: the production of beer, wine and liquor from alcohol, converting it to a recipe. Now let's take the example of Coca Cola and examine a list of the possible ingredients in Coca Cola's "secret formula," as provided by Wikipedia.

- *30 lb (14 kg) sugar*
- *2 US gal (7.6 l; 1.7 imp gal) water*
- *1 US qt (950 ml) lime juice*
- *4 oz (110 g) citrate of caffeine*
- *2 oz (57 g) citric acid*
- *1 US fl oz (30 ml) extract of vanilla*
- *¾ US fl oz (22.18 ml) fluid extract of kola nut*
- *¾ US fl oz (22.18 ml) fluid extract of coca*

Let's say we change the recipe in two places and that we are changing the quantity not the ingredients. By making any change we create a new formula. We can see how this new formula would look. Let's imagine that this is a new product to be introduced by Coke to compete with Red Bull. (To all you Coke people out there, I am aware that Coke already offers energy drinks, but this is just a hypothetical example.) It will be marketed as "Coke Awake." Notice the change to the two ingredient quantities, which are bolded below:

- 30 lb (14 kg) sugar
- 2 US gal (7.6 l; 1.7 imp gal) water
- 1 US qt (950 ml) lime juice
- **6 oz (165 g)** citrate of caffeine
- **4 oz (114 g)** citric acid
- 1 US fl oz (30 ml) extract of vanilla
- ¾ US fl oz (22.18 ml) fluid extract of kola nut
- ¾ US fl oz (22.18 ml) fluid extract of coca

In process industry lingo I changed the formula, not the recipe. This is because the recipe is the ingredients list, **and the formula is the proportion and quantity of each ingredient to be added to make the final product**. To non-process manufacturing people, this revised formula is just another BOM to maintain. Have twenty formulas for your recipe? Great; that is 20 BOMs. ERP and the planning system take care of all unit-of-measure issues. The term recipe is often used interchangeably with the term BOM; however, it is most accurate to consider a new BOM as any variant of recipe or formula. In an effective BOM or recipe solution, it is very easy to create variants, and this will be demonstrated further on in this chapter.

The Differences Between BOMs and Recipes and Formulas

If one thinks of the standard discrete manufacturing BOM it begins with many parts and culminates in a single finished good item. This is shown in the graphic below:

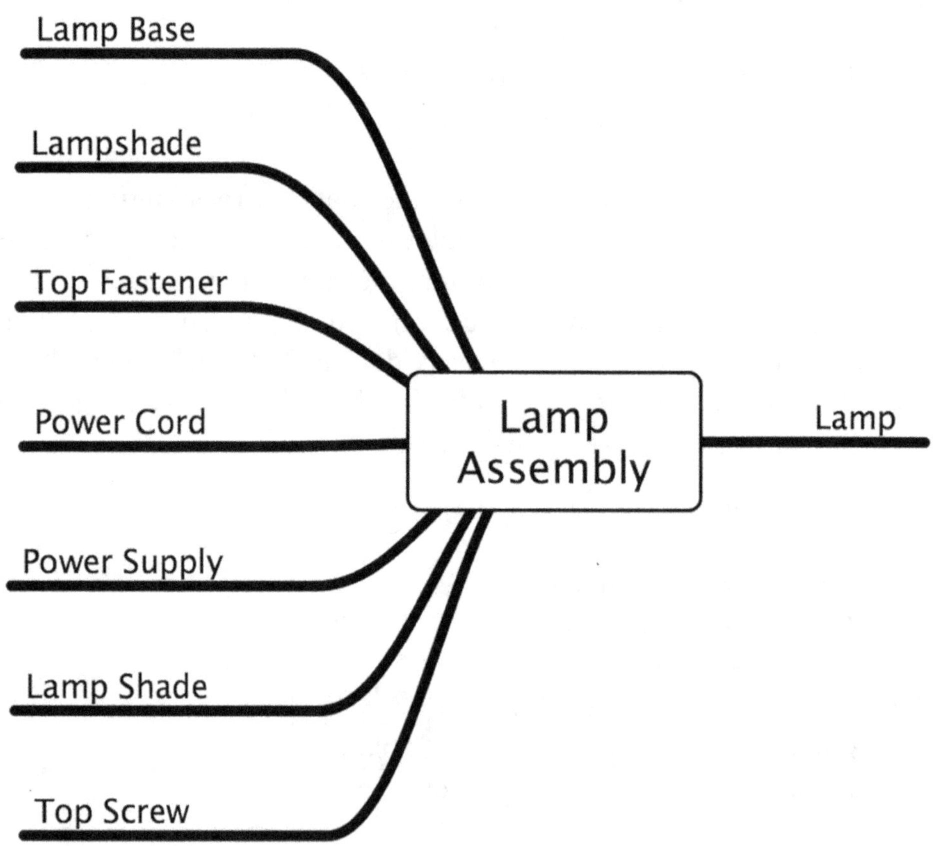

A typical discrete manufacturing ERP system, such as ERPNext, will have one field for the finished good, as can be seen in the following screen shot.

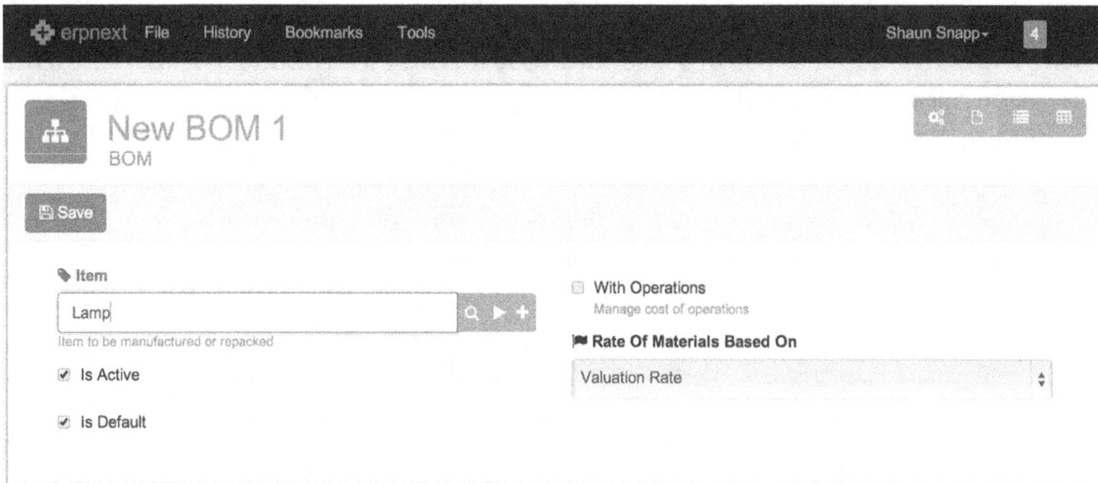

Here one can see that the BOM functionality in ERPNext is designed for discrete and repetitive manufacturing as one can only input a single finished good.

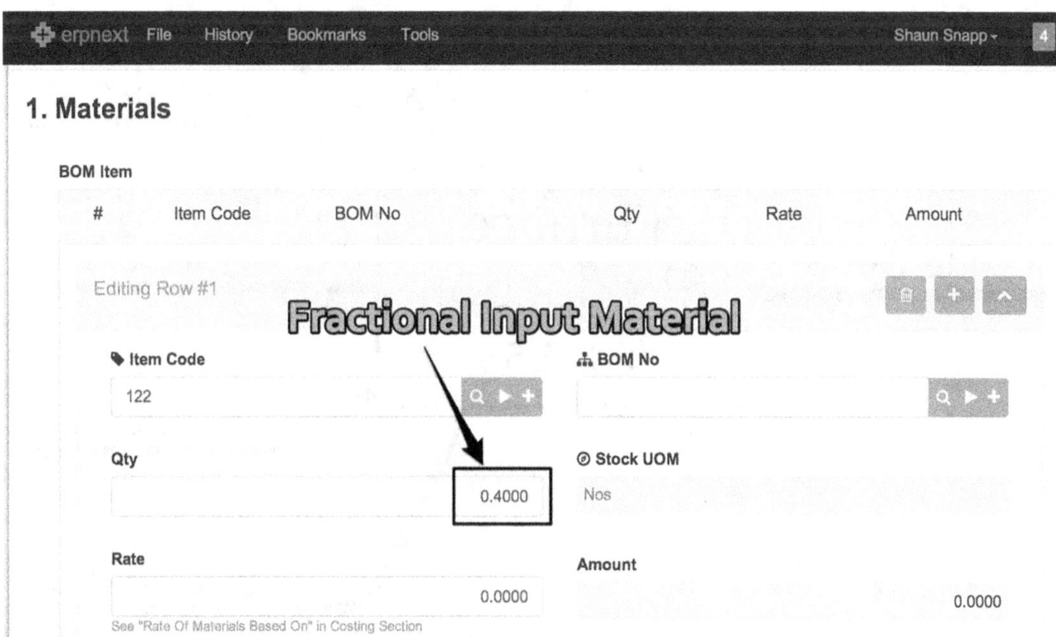

Although interestingly enough, ERPNext has the second functionality required to manage recipes, which is the ability to use a fraction as an input product on the BOM.

If the number of output/finished goods is limited to one, then this application will be unsuitable for most process manufacturing environments. The following quotation explains why this is the case.

> *"In food, pharmaceutical and chemical plants, many processes result in multiple outputs which looks more like an "X" structure. Take for example, a chicken disassembly process where the one chicken can produce 20–30 separate end products that can be sold or used for further processing. Another example is an oil refinery in which the raw material (crude oil) is refined into several types of product such as natural gas, rich oil, light cycle oil and decanted oil resembling a "V" structure."*[61]

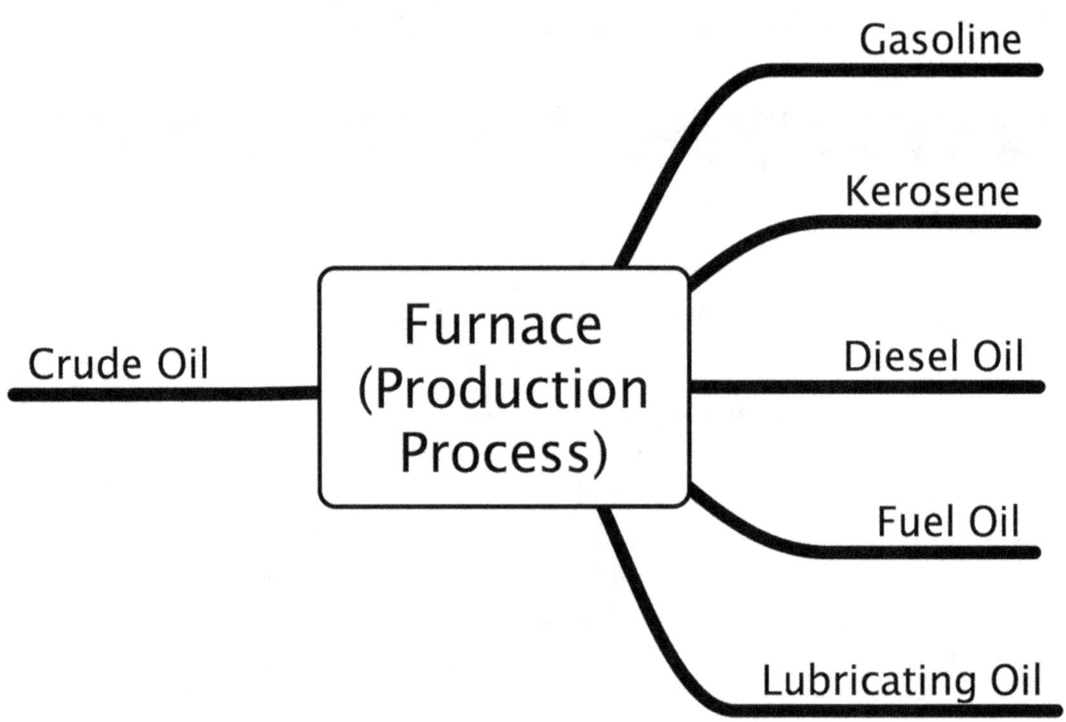

[61] Process Industry ERP Requirements

Recipes and Formulas

While the term "formula" is sometimes used interchangeably with the term "recipe," a formula and a recipe is not the same thing. A formula is, in fact, a subcomponent of a recipe. Formulas can be completely separate objects and a formula can be a production schedule all by itself—that is, it can stand separate from a recipe. In most cases a formula does not stand separately, and is combined as part of a subordinate object to a ***larger recipe, which contains multiple formulas***. It is common to have multiple formulas within a recipe because each formula relates to the manufacturing configuration of a semi-finished good. It is also quite common for semi-finished goods to have complex formulas of their own that must be processed as part of the in-process manufacturing steps. Let us now get into how recipes are used in SAP ERP (as an example of how recipes are used in ERP systems generally). But before we do that we need to explain the concept of a production version.

Process industry companies face the same challenges with respect to managing the product information as discrete and repetitive manufacturing companies when it comes to recipe and formula variants. Many companies undermine their capabilities by attempting to manage these objects in ERP systems—to which the SCM Focus Press book, *Bill of Materials in Excel, Planning, ERP and BMMS/PLM Software*, is focused on explaining why this is an ineffective approach. The fact that process industry finished goods can be disassembled while discrete and repetitive finished goods cannot has no interaction here, because finished products are not in fact disassembled—this is merely a way to differentiate between manufacturing categories. For instance, there can be a wide variety of formulas for the same recipe. All companies face the need for costing BOMs/recipes, which are driven by finance. For more on costing BOMs see the following link:

> http://www.scmfocus.com/productionplanningandscheduling/2011/11/10/the-problem-with-excessive-costing-bills-of-material/

The case must be made that it makes the most sense to treat recipes and formulas as if they are BOMs. Most material on this topic takes a different view, but in this chapter I hope to prove to you that there is no difference between the recipe, formula and the BOM aside from nomenclature, and that by thinking of them as different, one will tend to place unnecessary limitations on how you manage them.

The paper *What Makes Process Process?* makes the excellent point that the classic BOM tends to assume discrete or repetitive manufacturing. In the discrete or repetitive BOM, many parts are combined to make a single item. However, this is only one configuration. For instance with a poultry processor on chicken (the raw material), multiple finished goods result.[62] Frequently the BOMs for the process industry are multi-level and can be one to two levels deep or up to fifteen or more levels deep.[63]

Selecting the Recipe with Production Versions in SAP

In SAP ERP, a production version is the configuration object that determines which alternative BOM and master recipe are used together to produce a finished goods material. The production version functionality is part of the SAP PP (production planning) module. It is set up in SAP PP, but can also be used by the external planning system. Generally this data flows from the ERP system to the external planning system.

For one material several production versions frequently exist for various validity periods and lot size ranges. (In other areas of SAP these different versions would be called variants.) Production versions are used in both discrete manufacturing and process manufacturing.

---

[62] The paper *What Makes Process Process?* observes that, "*If your realities match the assumptions inherent in a Bill of Material, then many ERP systems can work for you. If your realities do not match—a system based upon a Bill of Material (even if it is called a "formula" in the sales pitch) needs very close inspection to see if it meets your requirements.*" I actually disagree with this statement, but not for the reason the author might assume at first blush. Perhaps the BOM functionality provided by ERP systems works **better** for discrete and repetitive than process industry, but the BOM functionality in an ERP system does not really work well for any type of manufacturer, and the ERP system should never be the system of record for BOM data.

[63] A multi-level BOM is explained well in the article, "Managing Multi-Level BOMs," by Arena Solutions. "*A multi-level BOM, also referred to as an indented BOM, depicts parent-child relationships and shows the hierarchical structure of assemblies and their related parts and components. A multi-level BOM is essentially a nested list whose parts or items are listed in two or more levels of detail to illustrate multiple assemblies within a product's BOM. In contrast, a single-level BOM depicts one level of children in an assembly and only the components needed to make the assembly are listed.*"

Creating and Reading Production Versions

Production versions are created with the following transaction or pathway within SAP ERP.

Logistics → Production Process → Master Data → Production Version

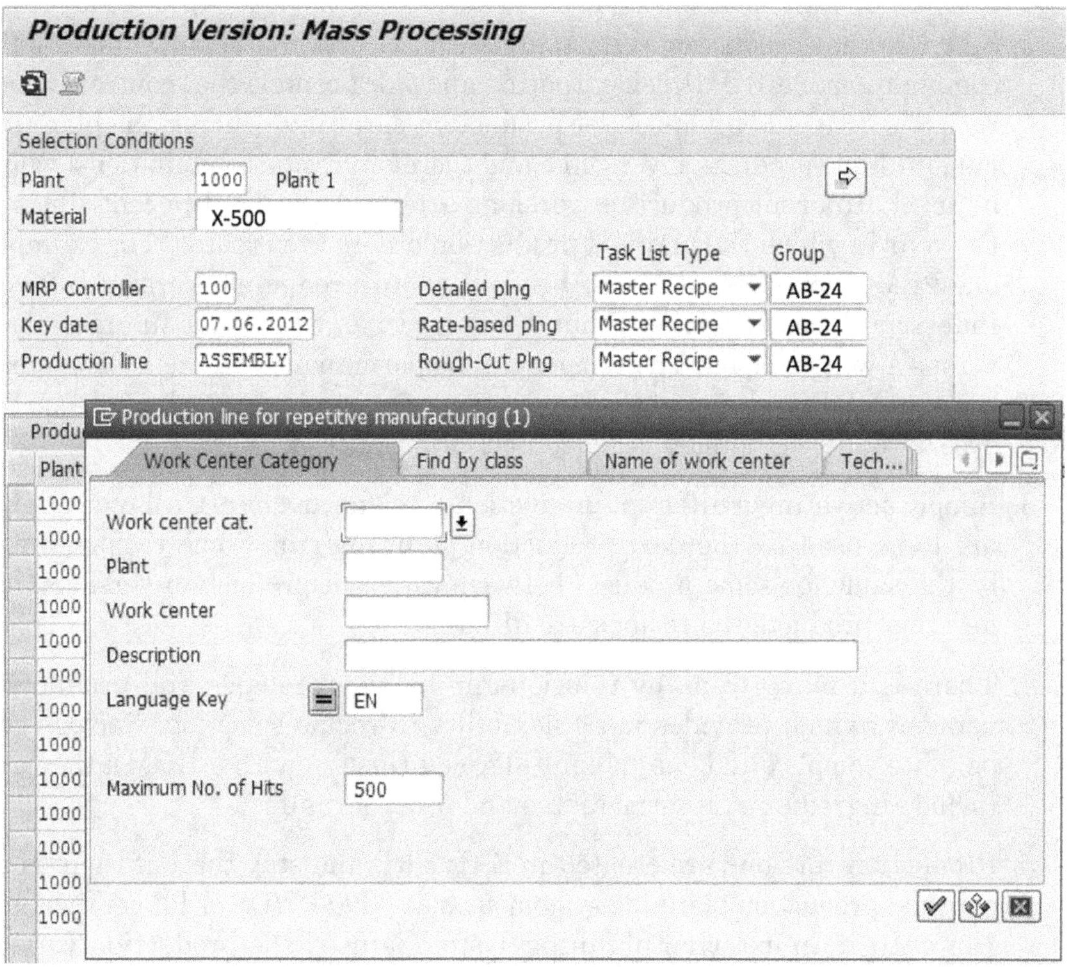

After you create a production version, the material components of the relevant alternative BOM are available in the recipe as a master list or can be created from the recipe. Within the production version you can assign a storage location to the material you want to produce.

Recipes in SAP ERP

The explanation of how recipe management works in SAP ERP is a bit involved. I think the explanation is easier to understand when read in a list format as follows:

1. Recipes can be altered in SAP ERP by copying them from older recipes and making changes to make new recipe variants (which would equate to a different formula of the recipe, but would also equate to a BOM copy).
2. SAP ERP has production version functionality (the production version is the combination of a BOM/recipe, routing and work center) that controls which recipe is activated and used for production planning. During one planning run, multiple production versions (and therefore recipe combinations) would be used; different production versions are used for the different times of the year in which the planned production orders are created. For example, one Production Version A may be identical to Production Version B for an industrial lubricant, except that it has a recipe, which is a summer blend versus a winter blend. The recipes can be permanently in the ERP system, but the validity period of the production version changes when one or the other production version is scheduled for production.
3. Simply activating a different production version can change all recipes that are to be used for the next production planning run. Some recipes would be the same for some products between different production versions, but for other products the recipes are different.
4. There is a many-to-many relationship between recipes and production versions, which provides more flexibility in recipe selection. Recipes can be quite complex and can be multi-leveled (just as with a BOM); they can include ingredients, intermediates and other formulas.
5. Production versions are created in SAP ERP and stay there if there is no external production planning system such as APO PP/DS or PlanetTogether. However, if an external planning system is used the production version is exported to this system during the normal master data update process between the systems.

Is Separate Nomenclature for Process Industry Constraint Master Data such as Recipes Necessary?

SAP has put forth a great deal of effort in creating a new area within their software for managing recipes, but is this actually necessary? A look at other vendors can be instructive. Unlike SAP, the vendor PlanetTogether does not use separate nomenclature in this area. When I first searched through the PlanetTogether training manuals, I did not find a single use of the term "recipe." I immediately knew that I was missing something because planning for process industry manufacturing is a strength of PlanetTogether. Shortly after, I spoke with John Cameron, a PlanetTogether Application Consultant. He explained to me that PlanetTogether does not use a separate nomenclature to define multi-level BOMs, routings and resources when applied to process manufacturing.

Upon further reflection I concluded that a separate terminology and configuration for these master data objects, although standard in process industry manufacturing, is unnecessary because the concepts and the way in which they work in the software are the same (as demonstrated by my "Coke Awake" example).

But this is not to say that process manufacturing BOMs can have a **great deal** of variability compared to BOMs, and that recipes have some different functionality requirements—such as the necessity for partial quantities as was shown in the "process BOM" example with Demand Works Smoothie, but how this is managed is a function of the capability of the application, **not what the BOM/Recipe is called**. Another example of this is how the BOM is represented in Demand Works Smoothie. It not only has fractional factors but it has the ability to enter a single input product and make multiple output products, and could assign multiple input products for multiple output products. Therefore, while the Demand Works Smoothie user manual does not directly address recipes, nor does the BOM table change for recipes, because of these characteristics, Smoothie is fully recipe capable. All of that, without any nomenclature changes, and without marketing documentation designed to lure the process industry manufacturing companies.

This consultant and I both agreed that the main benefit to changing the nomenclature for these software objects is that it matches the nomenclature used by clients and is a signal to potential clients that you, as a software vendor, can perform process manufacturing planning. However, there can also be the use of something called "specification blending," which does not fit a recipe. Once one is convinced of the applicability of the BOM to both process industries as well as discrete and repetitive, interesting options open up.

The Need for an Application to Manage your Recipe/Formula/BOM

The SCM Focus Press book, *Bill of Materials in Excel, Planning, ERP and BMMS/PLM Software* provides ample evidence, which is laid out in a complete form that every company requires an application for managing their BOM/recipe/formulas. This book was focused on the BOM and the application Arena Solutions, but the exact same principles apply for recipe/formula management. The use of these types of systems allows the company using the application to more efficiently perform the following activities:

1. Improve design efficiency.
2. Allows for more simulation and trial and error—promoting creativity in the design process—all without the necessity of creating permanent records in process of developing new designs.
3. Improves internal understanding and coordination on recipes—some applications like Hamilton Grant allow for manufacturer's instructions and trial notes to be stored.
4. Groups working in different geographies can more efficiently work as teams and coordinate their efforts. Workers who are not in the office, but may need to provide review input can access the system remotely to provide critical and timely input.
5. Improves the speed of recipe/formula development.
6. Allows companies to better manage their product costs.
7. Allows companies to more efficiently integrate with suppliers by sharing and providing access to either the entire recipe/formula or just to portions of the recipe/formula. In fact, these types of systems double as supplier management systems as well as product data management systems. Once product data is in a highly organized and sharable form, it has numerous uses.

8. Related to the previous point, allows the company to work with more suppliers as the costs of interacting with any one supplier is lower.
9. Keeps all of the ingredient, packaging, recipe, process and quality information in one place.
10. Improves recipe/formula re-usability.
11. Improves compliance with legislative standards (which vary greatly by industry—but are quite important in food and beverage and pharmaceuticals).
12. Simulate the effect on important indicators during potential redesigns.

At our companion site, Software Decisions[64], we estimate that BOM/recipe/formula management applications have the highest return on investment of any enterprise software category that we track. A main reason for this is so many parts of the business rely upon product information and therefore can benefit from a single application. Design and engineering, supply chain, finance, supplier relations—all of these sides of the business benefit. Furthermore, each side of the business, as well as external parties, are better integrated through the use of this type of application. Here is an example from one vendor of how their recipe management software works:

> *"Recipe management can even create least-cost recipes from scratch. You can reuse your knowledge by using templates. These templates define the ingredients that can be used and the constraints on nutritional content and ingredient quantities (even for rework).* — Adifo Software

Recipe Management with Hamilton Grant

Most of the recipe management applications that currently exist are directed toward the food and beverage industry. This is thought to be because of their stringent requirements, many of which were described previously. Change is a constant feature of food and beverage product information as these companies will frequently reformulate their recipes for every production batch.[65] Hamilton Grant's

[64] Software Decisions can be found at http://www.softwaredecisions.org

[65] http://www.hamilton-grant.com/The-HG-Solution/HG-Recipe-Management/Formula-Optimizer

software can greatly reduce the effort involved in doing this because they have functionality that can optimize the reformulation process—such that the objectives are set for the application and then the application will meet these objectives at the lowest cost. Therefore, the **optimizer** determines what the new formula will be.

Recipe management software can be found outside of the food industry, but the selection quickly narrows. The differentiation between a BOM and a recipe is most easily seen inside of an application like Hamilton Grant because their application provides the type of information that show ingredient aggregation as the food designers are building a new recipe, they must know how various indicators are changing and whether they have exceeded the specification as food must increasingly meet certain nutritional and other various compositional targets.

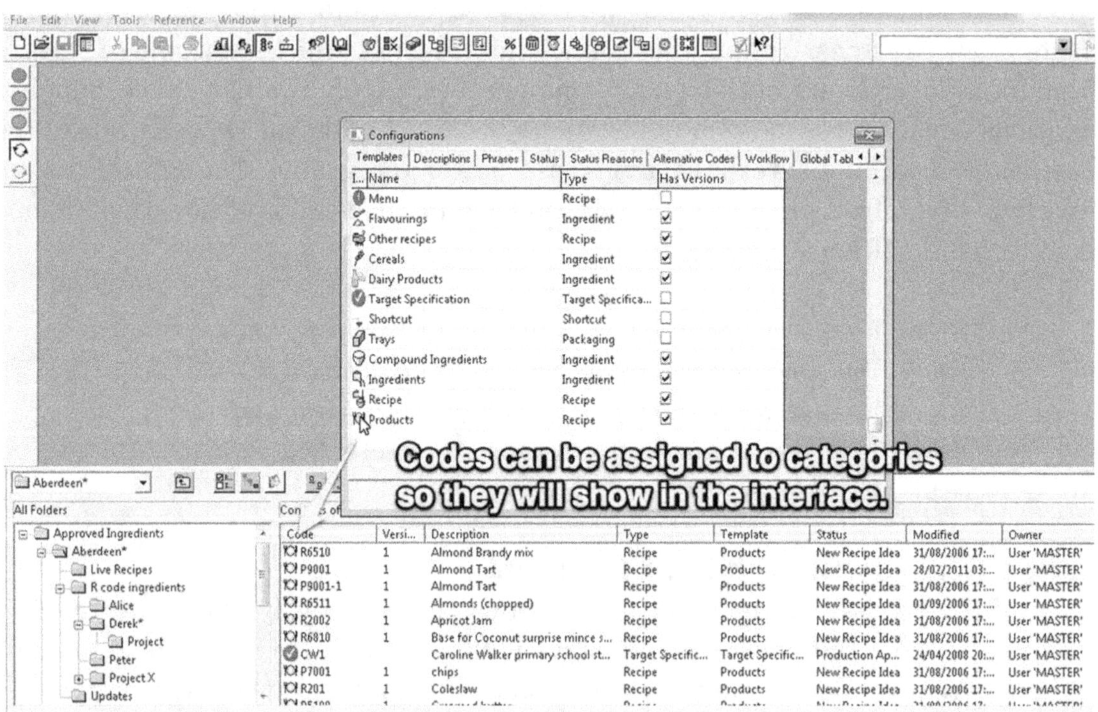

To begin, the icons that show up in the application can be set by the system configurator.

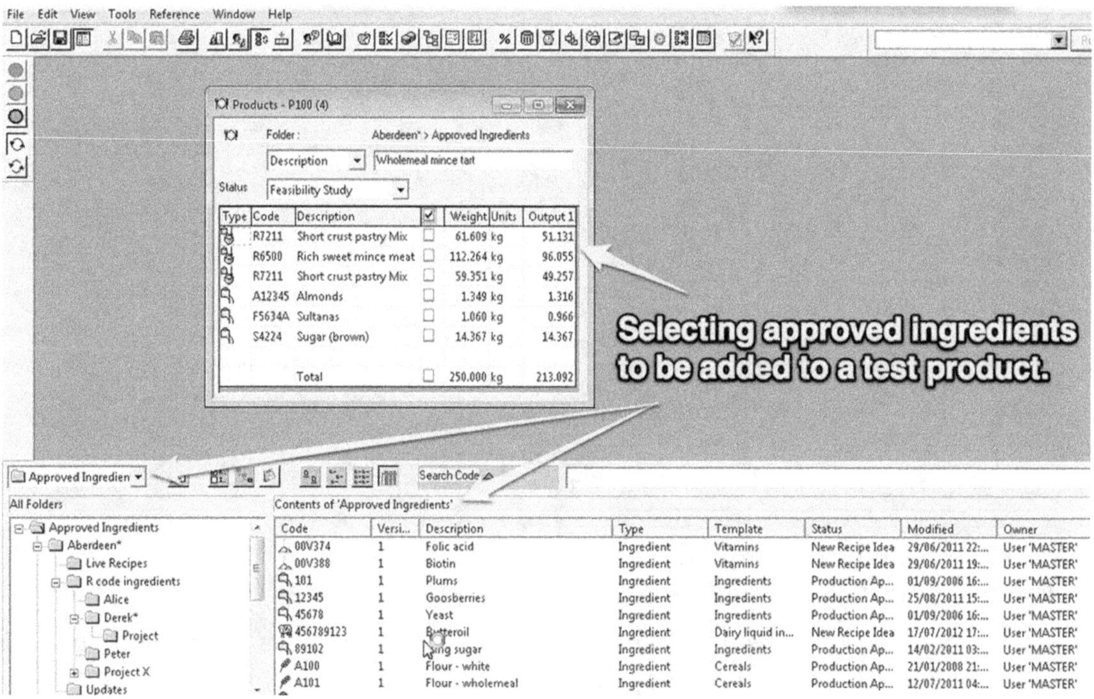

As with all good recipe/BOM management applications, designers use them intensively, and many of the iterations that they create will never find their way to the production planning system. However, the sub-portion of the overall recipes that are created can be sent to ERP systems (and from there to planning systems) in a complete form. Recipe/ BOM management systems allow for the easy duplication of recipes by using a template prior to developing a new design or product. In the case above, a food designer is adding a series of ingredients (which are only approved ingredients) that will make up a new mincemeat pie. As the food designer adds ingredients, the characteristics of the product are totaled below. The Hamilton Grant Recipe Management application can hold a wide variety of ingredient characteristic, but only a few are shown in this view. A right mouse-click allows the designers to add and delete the characteristics that will show in the user interface so they can see the summation of the ones that interest them.

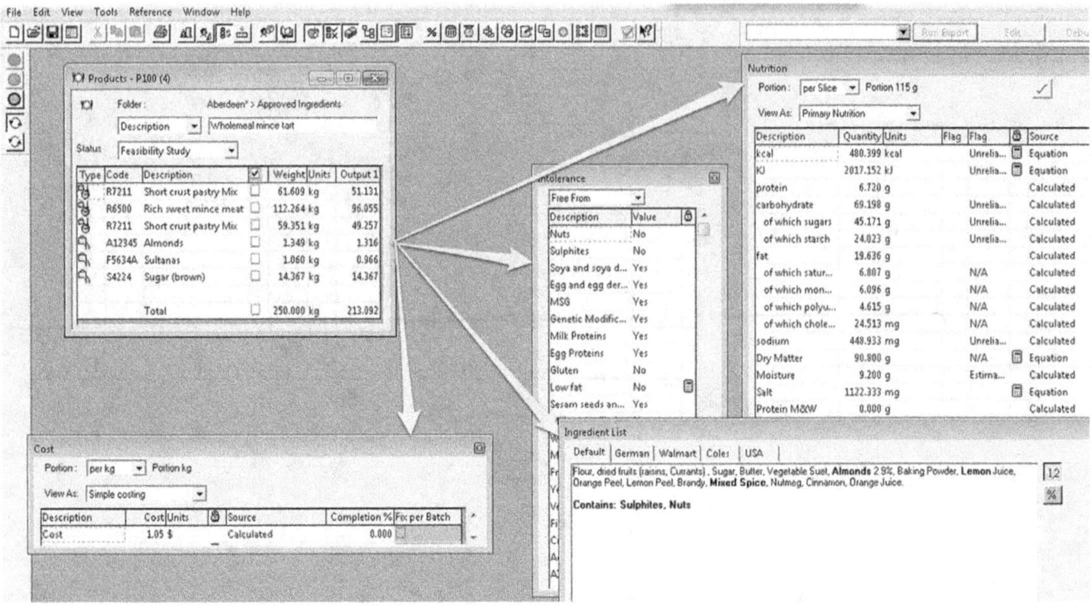

New windows can be opened that show more a detailed analysis of the new product that is being designed. In this screen shot you can see the costs, intolerance, nutritional content and the ingredient list. These perspectives on the product help the designer make adjustments to the products that meet the design objectives. For instance, food companies very much design their products in a way that result in a marketable food label. By viewing the label each time a new ingredient is added or removed, the food designer can make the adjustments to result in the desired food label.

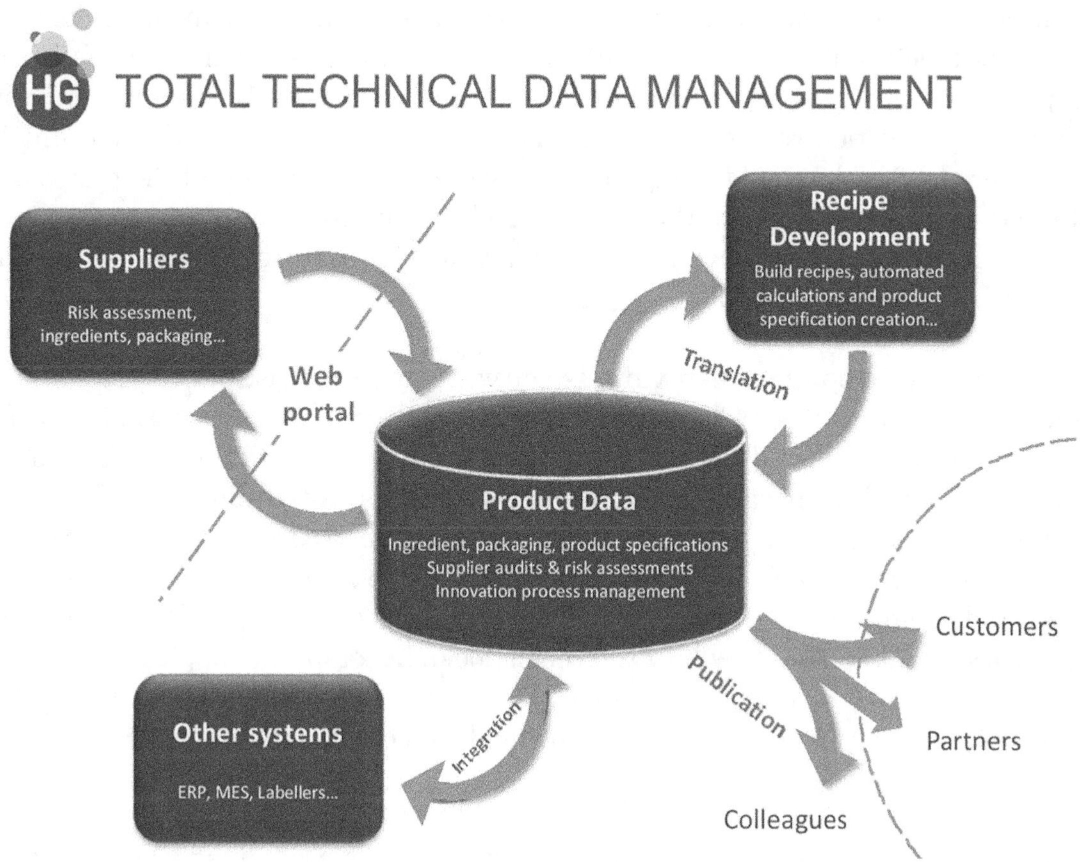

The strength of these types of applications is that the information that is created in them is shared with many other applications. They serve as a hub for product data— a far superior approach to the poor design of using ERP systems as the hub of product data. In fact, the ERP-centered approach to being the product data hub has been attempted extensively and has been proven to ***not work well.*** *Even so it is still the dominant approach employed in companies. When companies follow this approach, ERP systems end up being a bottleneck to the implementation of other systems that rely upon product information.*

Hamilton Grant's software is used first by product design before anyone from supply chain touches the recipe or formula. Product design makes numerous revisions (copies of new recipes and formulas) before a production recipe is created. Even after the production recipe is created, there are still more copies that are made as the product is changed through its lifecycle or as other copies of the recipe are created. Hamilton Grant's ability to manage large numbers of revisions and store them all—and to allow anyone with the access rights to see them—is a major strength of the application.

Once the interaction between various software applications and the recipe are understood, the next step is to learn how to leverage applications that use the recipe. I believe that individuals have difficulty understanding the recipe in its comprehensive form because existing literature does **not include application screen shots to show how the recipe looks in different applications**. While some product manuals produced by the recipe management vendors come close and are quite good, the manuals focus exclusively on a particular vendor's application and do not show how the recipe looks in competing applications. As a result, few people receive the benefit of a comprehensive education in this area, which is a primary reason why the recipe tends to be so poorly managed generally.

Conclusion

A BOM is simply a list of input items that connect in a predefined way to an output item or items in certain proportions. The BOM is one of the most constantly changing master data elements.

While the term "formula" is sometimes used interchangeably with the term "recipe," a formula and a recipe are not the same thing. A formula is, in fact, a subcomponent of a recipe. For instance, it is common to have multiple formulas within a recipe because each formula relates to the manufacturing configuration of a semi-finished good.

Process industry manufacturing companies face the same challenges with respect to managing the product information as discrete and repetitive manufacturing companies when it comes to recipe and formula variant management. Unlike SAP, the software vendors PlanetTogether and Demand Works do not use the process industry manufacturing nomenclature in this area, but both are capable of process industry planning because they have designed their functionality to do so. The ability to manage BOM/recipe information in a way congruent with process industry manufacturing needs is a main component of what makes any application process industry capable. While it is certainly true that process-manufacturing recipes can have a great deal of variability, how this is managed is a function of the capability of the application, not what the BOM is called.

In ERP, planning systems, MES process control, the two major differentiators from BOM functionality is the ability to have fractional inputs, and the ability to have multiple output products. However, the differentiation between a BOM and a recipe has other dimensions that are most easily understood by reviewing a recipe management application like Hamilton Grant. This is because their application provides the type of information that show ingredient aggregation as the food designers are building a new recipe. Food designers must know how various indicators are changing and whether they are meeting the predetermined specification. This is due to the fact that food must meet certain nutritional and other various compositional targets. This is not the focus of BOM management vendors, for instance like Arena Solutions.

It should also be considered that software purchase decisions are most often shared with design and engineering, as they set up the recipes/formulas initially **before** they become production-ready and before anyone in manufacturing actually sees them. In fact, the recipes and formulas that manufacturing sees will most likely be after the recipe and formula have gone through a number of iterations. Hamilton Grant's software is used first by product design before anyone from supply chain touches the recipe or formula. Even after the production recipe is created, there are still more copies that are made as the product is changed through its lifecycle or as other copies of the recipe are created.

CHAPTER 6

Overlapping, Leading and Lagging Operations

The Importance of Timing in Process Industries

Of all of the different types of production environments, process industry is the most sensitive to timing. If one were to stop work without notice at a discrete or repetitive production plant, the work could be picked up the next day, and the work in process would in the vast majority of cases not have changed. It would be undesirable of course, because there would be lost productivity, but it is quite different from what happens if production were to be halted without notice in a process industry plant. While all manufacturing categories want to control the time element so as to maximize productivity, process manufacturing is the most time sensitive because time actually changes the characteristics of the input products; work in process, and output product. Of course one immediately thinks of the food industry in this regard, because of the high rate of perishability of food products. For instance, if vats of milk were to be left overnight in various intermediate states and the plant were to begin production the next day, it's easy to see how most of this cheese in-process would be unusable. This is true for both the input products and the intermediate and output products. In milk production for example, when milk trucks arrive at a facility, they are drained and held in tanks with the intent of processing all of the milk from the tanks as quickly as possible.

One of the most basic types of time-dependent processes is the aging process, as this cheese is being aged. Aging is common in alcoholic beverages as well.

Whiskey is often rated based upon how long it has been aged. Aging alters the whiskey, and also makes it pick up flavors from the barrel. Aging at the end of the production process is one of the simplest time-dependent features to manage. Things become much more sensitive when they are in the middle of the manufacturing process.

The issue of timing is consistent across different categories in process industries. Even mixing too long can have negative consequences, as the quotation below explains.

> *"There are a number of situations where excessive mixing in some form or other is not only wasteful but is counterproductive. Thus in the mixing of biological materials, excessive inputs or impeller speeds may damage mammalian cells of mycelia in agitated bioreactors. Some polymer solutions whose desired rheological characteristics result from structured molecular forms may have been broken down and not reformed."* — Mixing in the Process Industry

The Importance of Temperature in Process Industries

Directly related to timing is the temperature component of many forms of process industry manufacturing. As the quote below describes, in some process industry categories, the temperature is critical all along the different manufacturing processes.

> *"Almost every process in the food industry involves thermal effects such as heating, cooling or phase transition. The thermal properties of foods are therefore of considerable relevance."* — Food Process Engineering and Technology

Paper, chemicals, metals, power generation, food, beverages, glass and pharmaceuticals are all examples of process industries that use temperature intensively in the manufacturing process. Materials must be kept at a certain temperature for a specific amount of time. Often they must be combined with other materials after this temperature treatment.

- In the petroleum refining area, when crude oil is heated and distilled, different products (such as butane, gasoline, naphtha, kerosene, gas-oil and residue) are separated by boiling-point temperature through the distillation curve.
- Glass production requires the input products being placed into a glass furnace for the initial heat treatment, and then repeatedly heated and cooled in ovens.
- Power generation is highly dependent upon temperature, as is the pulp treatment process in paper manufacturing.

Timing and heat are good examples of properties of process industry manufacturing—which leads to the general topic of leading and lagging operations.

Leading and Lagging Operations in Software

A major requirement in scheduling for process industry manufacturing is leading, lagging and overlapping scheduling capabilities. Operations are not simply scheduled one after another, nor are concurrent operations scheduled to begin at the same time.[66] Leading, lagging and overlapping operations determine how soon a successor operation can be started before the predecessor operation completes. While a perfect definition as to what differentiates leading and lagging operations from overlapping operations has not been agreed upon, the phrase "overlapping" tends to apply where one process is dependent on the completion of some portion of the prior or follow-on step. In refineries, most of the flow is continuous from one operation to the next, but there can be intermediate tankage, which is then processed by itself.

For PlanetTogether, leading and lagging and overlapping operations are very easy to configure. One can create an overlapping or lagging schedule of jobs as part of its normal functionality, and the vendor has this functionality live at most of its accounts.[67]

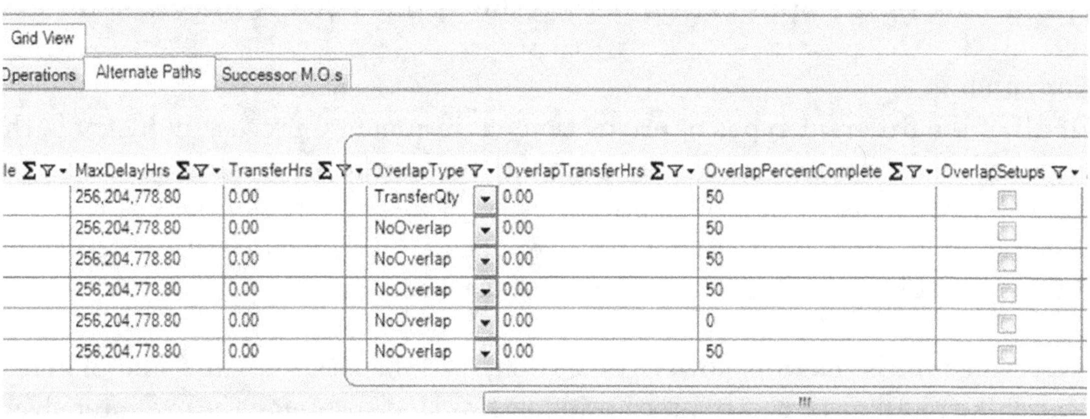

This screen shot shows the configuration screen to lag or overlap operations.

The overlap functionality is so detailed in PlanetTogether that it can be set on the basis of ***a percentage of total cycle capacity***. So, for example, after 35 percent of one process is complete, the next operation can begin, making the overlapping functionality ***highly specific and furthermore easy to maintain***. If the lag is

[66] As a side note, I had been using the term "leading and lagging operations" with clients and in articles for some time. When doing research for this book, I was surprised to find that the term was not used that commonly. However, the term "overlapping operations" is also not found often on the web. I am unsure why this should be the case, as this principle is so important.

[67] Dave Biddle, Senior Application Consultant, PlanetTogether

set to zero percent, the operations are sequential. If the lag is set to 99 percent, then the operations are very close but not quite concurrent. Some applications require that the operations be set as FS (Finish Start), SS (Start, Start) rather than applying a specific percentage lag, which I personally find much less clear and less specific than providing a percentage.

I have not performed a survey of all currently available production planning applications; however, I believe this functionality, and particularly its **ease** of configuration as is demonstrated here in PlanetTogether, is rare. Any application that is to be used for scheduling in-process industries should be checked for this functionality during the software selection process. This percentage lag is so critical to process industry manufacturing that for me it would be a "have to have." Furthermore, production schedulers should have the opportunity to **lag operations in the application before the application is purchased**.

Conclusion

Of all of the different types of production environments, process industry is the most sensitive to timing. While all manufacturing categories want to control the time element so as to maximize productivity, process manufacturing is the most time sensitive because time actually changes the characteristics of the input products; work in process, and output product.

Paper, chemicals, metals, power generation, food, beverages, glass and pharmaceuticals are all examples of process industries that use temperature intensively in the manufacturing process. Materials must be kept at a certain temperature for a specific amount of time. Often they must be combined with other materials after this temperature treatment.

A major requirement in scheduling for process industry manufacturing is leading, lagging and overlapping scheduling capabilities. Operations are not simply scheduled one after another, nor are concurrent operations scheduled to begin at the same time. Timing and heat are good examples of properties of process industry manufacturing—which leads to the general topic of leading and lagging operations. Any application that is to be used for scheduling in-process industries should be checked for this functionality during the software selection process.

CHAPTER 7

Mixing/Blending/Pooling Operations

"Mixing" is the combining of different input products in order to produce a uniform output product. Mixing operations occur frequently in chemical processing and in food processing. As stated in the book *Chemical Manufacturing: The Process of Mixing*:

> *"The goal of the mixing process is to achieve a fully randomized sample."*

Mixing has a wide range of purposes, one being to provide products on a wide range of specifications. For instance, at a plant that makes industrial lubricants there may be hundreds of lubricant specifications. The plant meets these specifications through mixing/blending. The mixing operation must use the correct input products mixed at the correct temperature for the correct amount of time. In lubricants, the output product must meet a long list of performance characteristics, as is listed below:

1. Viscosity
2. Oxidation Resistance

3. Foaming Resistance
4. Emulsion Resistance
5. Corrosion Resistance
6. Deposit Resistance
7. Friction Resistance[68]

Mixing durations can span from seconds to hours. Sometimes mixing involves multiple liquids (generally the easiest type of mixing operations). But there is also the mixing of solids with liquids—and the mixing of solids and gasses as a way to imbue the solid with the gas, as in the case with foam and metals—as well as the mixing of solids. Mixing is part of the production process for many obvious products such as paint, but also less obvious products such as glass. Normally glass is mixed in glass vats by combining sand, soda ash, limestone and cullet ("crushed glass of the same type that is being produced").[69] While mixing can be relatively simple, complexity can increase rapidly depending upon what is being mixed. In many cases, much more is happening at the same time as mixing is occurring. For instance, mixing is often combined with heating, agitation, and chemical treatment. Homogeneity of the sample is only one example of the output of the mixing operations.

Mixing is one of the most important quality steps in manufacturing operations that include mixing. Often the quality and consistency of the mix determines a good part of the quality of the finished good. There are a number of specifications that the mixture must meet before it is taken to the next manufacturing process.

Most mixing operations are performed in batch, but not all of them. Some mixing operations are continuous; in this case materials are added continuously to the mixer and fed continuously to the next step in the manufacturing process.

[68] Petroleum Refining in Nontechnical Language
[69] *Chemical Manufacturing: The Process of Mixing*

Mixing is a major part of cheese production. The picture above is the mixing of curds and whey until they reach the desired temperature and firmness.[70] *After this point the whey is drained, leaving the curd.*[71]

When combined with packaging after the mix and some material preparation prior to the mixing, paint mixing comprises most of the manufacturing operation (the rest being packaging). More frequently, mixing is not the primary portion of the manufacturing process; normally it takes place at the beginning or at least very early in the manufacturing process.

While I have referred to the overall category of mixing/blending/pooling as simply "mixing," in fact there are differences between the three. Mixing is the most general term, but blending implies a gentler process than mixing. Pooling is also

[70] Steps in Making Cheese

[71] Whey is a co-product of cheese manufacturing, because it is a sellable product, however it was not always the case. Before the market for exercise drinks and related uses of protein, whey was considered a by-product.

different. Up to this point I have described mixing as a single operation, but this is not always the case. The complexity of mixing operations is explained by the problem known as pooling. Pooling is described well in the paper *The Pooling Problem.*

> *"Given a list of available suppliers (inputs) with raw materials containing known specifications (specs), what is the minimum cost way of mixing these materials in intermediate tanks (pools) so as to meet the demand and spec requirements at multiple final blends (outputs)?"*

The pooling problem is also referred to as "feedstock optimization." It's a common issue in the oil refining business. Here the company buys titanium dioxide ore from multiple sources and optimizes the mix of ores for the best price and yield while staying within processing limits for arsenic and nickel.

There's also "specification blending" in which two or more tanks of materials are mixed together in order to get to a saleable finished product that is on spec or not over spec so as to not have "product giveaway." To make on-spec gasoline, gasoline blending can involve more than a dozen tanks—preferably at the cheapest cost.

This is one of the features of process industries: you can take the discrete industry equivalent of six blue cars and combine them with six red cars to get twelve purple cars. Different chemicals can be combined to make totally different finished products.[72]

Using Software to Manage the Complexities of Mixing

As is discussed throughout this book, most process industry manufacturing is simply too complex and specialized for the software put out by most software vendors. Mixing operations can be simple or complex. For instance, polymer mixing—which results in plastics—can be quite complex. However, if the mixing problem is not complex, mixing scheduling can be performed quite easily within a spreadsheet. I have included the following example of a mixing scheduling spreadsheet to demonstrate this feature of mixing.

---

[72] Dave Biddle, Senior Application Consultant, PlanetTogether

System Output

Planner Added Values
1. Color Sequence Determined by Lookup Table
2. Production Capacity

Order Number	Resource	Start Date	Product	Product Short Description	Total Qty	Rec/ReqQty	COLSEQ	
190	ABC	8/17/11	119791	White Paint Interior	2,100	-	1	
201	ABC	8/18/11	122327	White Paint Gloss	4,000	-	1	
198	ABC	8/17/11	119788	White Paint Flat	5,000	-	1	
210	ABC	8/18/11	123159	Yellow Paint Gloss	7,500	1,000	2	
205	ABC	8/18/11	121750	Yellow Paint Flat	10,000	4,000	2	
185	ABC	8/17/11	119311	Blue Paint Flat	8,000	-	3	
189	ABC	8/17/11	116151	Black Paint Gloss	5,000	-	4	
					41,600	5,000		40,000 Production Capacity
							6	3,400 Capacity Available

The scheduling process requires several steps, which can be emulated in something as simple as a spreadsheet, using Excel's sort function.

1. Group orders that have close start dates.
2. Stop grouping the production orders once they approach the capacity for the day. This is shown by the following formula: Total Quantity for all **Production Orders – Rec/ReqQty < Production Capacity.** (These colors match the variables in the screen shot. Rec/ReqQty is the amount of the production order that has already been produced.)
3. Run a VLOOKUP formula to assign a Color Sequence value to each prospective order.
4. Then run a sortation, just on the day that is scheduled.

Conclusion

"Mixing" is the combining of different input products in order to produce a uniform output product. Mixing durations can span from seconds to hours. Sometimes mixing involves multiple liquids (generally the easiest type of mixing operations). But there is also the mixing of solids with liquids—and the mixing of solids and gasses as a way to imbue the solid with the gas, such as is the case with foam and metals—as well as the mixing of solids. Mixing is one of the most important quality steps in manufacturing operations that include mixing. Mixing operations can be simple or complex. For instance, polymer mixing—which results in plastics—can be quite complex. However, if the mixing problem is not complex, mixing scheduling can be performed within a spreadsheet.

CHAPTER 8

Changeovers and Cleaning

"Changeover" is a very general term that can mean a number of different things. Changeovers can typically be divided into cleanup, setup and startup. In process industry it is common to see the terms "setup" and "changeover" used interchangeably. SAP PP/DS uses the term "Setup Matrix" and PlanetTogether uses the terms "setup table" and "setup code." However, as has been explained, the setup is ***only one component of the changeover time***.

In process industry manufacturing, a big part of the changeover process is the cleanup or cleaning. Cleaning is used between the production of different products to prevent contamination between products, as well as after a number of hours have passed. The main point of cleanup is to both enable the next processing operation to operate smoothly and to prevent the previously produced item from polluting the next product to be produced. PlanetTogether has the following comment on how cleaning can be scheduled in their application.

> *"Equipment cleanouts are required to maintain hygienic conditions and prevent cross-contamination between products. Cleanouts are often scheduled during nonproductive hours*

(weekends for example) when day laborers are not present except for the personnel who perform the cleanouts. In this situation, PlanetTogether can be set up to always schedule cleanouts on weekends or during other specific time frames. If there are "minor cleanouts" that can be performed during the week for example and "major cleanouts" that should only occur on weekends then this can also be handled automatically.

In the event that cleanouts can be scheduled during productive times, Lookup Tables are defined to specify cleanout times required when switching between products with different attributes. For example, a cleanout when switching from beef to chicken may be longer than if switching between two different beef grades. The system will automatically schedule the appropriate amount of cleanout time, even when a drag-and-drop reschedule is performed to change the sequence. Additionally, the Optimization algorithm can be set to avoid excessive changeovers and thus maximize production output automatically and prevent cross contamination/allergens."

The Emphasis on Changeovers

At first blush, changeovers would only seem to apply to batch manufacturing and not to continuous manufacturing, which is one reason why manufacturing in a continuous fashion is more efficient than manufacturing in batch, as all changeover time is lost production time. However, changeovers do take place in continuous manufacturing. For example, changing pressures and temperatures can alter conditions of a continuous operation in order to transition to a different product quality. The material that is produced while the transition is occurring is neither "fish nor fowl" and matches neither the specification of the product being transitioned from or the one being transitioned to. Sometimes these intermediate transition products are siphoned off from the pipeline carrying the material and reprocessed at the same plant or sold to a third party jobber who separates it into usable material.[73]

[73] David Biddle, Senior Application Consultant, PlanetTogether

Changeovers are a frequent topic of conversation among management, who continually want to reduce the time it takes to change over. Often the term "flexible manufacturing" is used—as if using a word changes the constraints on the plant floor. Consultants have sold projects based upon the idea of fast and often unattainable changeovers. From the production planning and scheduling perspective, the most important thing is that the reality of the changeover times (and sometimes costs) are reflected within the application. As with all information fed to the system—the more accurate the information, the better the plan. Reduced changeover time obviously means more time available for producing, but production planning and scheduling software does not improve the ability of a company to physically perform faster changeovers. How the changeovers are managed by the application is a very important part of what a production planning and scheduling application does. It would be even more important, but the data related to changeovers that the company maintains is almost always less than what is desired by those who set up the application.

Changeover values are added to the production planning and scheduling system so that software has visibility to the extra time and cost that it takes to perform a changeover, and so that sometimes the software can optimize the overall production schedule in order to minimize changeovers. The software has to account for when the finished product is required, when supplies are planned to arrive, the length of each operation in the manufacturing process, the time between the operations, the inventory that must be carried if finished goods stock is produced before its requirements date, etc. Some of these values are input and calculated by the supply planning system and some by the production planning and scheduling system.

Changeover Product Relationships

There are two basic ways to plan changeovers in scheduling applications.

1. Enter changeover values that are independent of the previous production run—essentially to use an average changeover time.
2. Have the relationship between the product that comes before the next product built into the changeover logic.

The first way is not representative of reality; the assumption with this type of setup is that there is no dependency between production runs, which of course is not true. For example:

- In paint processing, cleaning is often required if a darker color is run before a lighter color, but often not so if the lighter color is run prior to the darker color.
- If a white cheese is scheduled after another white cheese, considerably less set-up is required than if a colored cheese precedes a white cheese.
- If a product that shares identical packaging characteristics to the product that precedes it, the set-up time and cost is lower than if the product were different.

These are called dependent setups, and they are extremely important for production scheduling because they reflect the reality of the production floor. Therefore, companies that have these constraints move their production cycles to create runs sequentially from lighter to darker colors, or from similar to less similar products, even if this is not necessarily the order in which the products are demanded. Following this approach, the finished goods stock will stay longer in inventory and incur higher inventory carrying costs. Some cost optimizers promise the potential of being able to trade off all of these costs so that an optimized schedule is delivered. However, this occurs very rarely in practice. There are many issues interfering with the objective of an optimized schedule that accounts for setups, one of the most common being that most companies do not actually know the costs of everything they do. Secondly, in the vast majority of cost optimizer implementations, actual costs are not used. The common problems with cost optimization, both in supply planning and production planning are discussed in the articles below:

http://www.scmfocus.com/supplyplanning/2011/07/09/what-is-your-supply-planning-optimizer-optimizing/

http://www.scmfocus.com/supplyplanning/2011/07/24/why-the-cost-of-missed-demand-must-always-be-set-so-much-higher-than-the-profit-margin/

Regardless of how many companies run into problems with cost optimization, not only do the sales people at vendors continue to pitch the same vision, but the vendors' consulting divisions and consulting companies continue to implement the same problematic approaches to the implementation.

> http://www.scmfocus.com/scmhistory/2010/07/the-rise-and-decline-for-supply-chain-optimization/

Changeovers are typically modeled in planning applications using a changeover table. As changeover tables introduce more complexity, to begin I want to communicate the concept the simplest way possible by showing a matrix I created in Excel. This matrix will demonstrate what the changeover table is attempting to accomplish.

Changeover Matrix/Table

From Product	To Product	Changeover Time (In Hours)	Changeover Cost (in Dollars)
White Paint	Yellow Paint	1.5	$ 550.00
Yellow Paint	White Paint	3.0	$ 1,100.00
White Paint	Dark Blue Paint	1.5	$ 550.00
Dark Blue Paint	White Paint	4.0	$ 1,466.67
Yellow Paint	Dark Blue Paint	2.0	$ 733.33
Dark Blue Paint	Yellow Paint	4.0	$ 1,466.67

Here we can see that moving from dark color paints takes longer and costs more than moving from light color paints. We also see that paint colors are on a gradient with respect to how long it takes to perform a changeover.

Let's observe the total changeover time and costs incurred depending upon the sequence that we create in the production schedule.

Paint Production Sequence

Product 1	Product 2	Product 3	Total Changeover Time (In Hours)	Total Changeover Cost (in Dollars)
White	Yellow	Dark Blue	3.5	$ 1,283.33
White	Dark Blue	Yellow	5.5	$ 2,016.67
Yellow	White	Dark Blue	4.5	$ 1,650.00
Yellow	Dark Blue	White	6.0	$ 2,200.00
Dark Blue	White	Yellow	5.5	$ 2,016.67
Dark Blue	Yellow	White	7.0	$ 2,566.67

Under the assumptions from the previous table, notice how much money and time can be saved by choosing the sequence white—yellow—dark blue. This is a small sample; if the production goes back to white after dark blue is produced (instead of to another dark color), the high costs and changeover time for that changeover must be incurred.

Sometimes the issue that drives changeovers is the amount of time required to clean colors; at other times the issue can be related to food safety. These types of restrictions are not only cost-based, but can be strictly prohibited by the system as explained in the following quotation:

> *"A common goal in Food Process scheduling is to avoid certain products from running in sequence on the same equipment, thus preventing allergens from cross-contaminating other products. For example, a mixture containing peanuts followed by another mixture that does not contain peanuts could lead to traces of peanuts in the second mixture. There are several strategies for avoiding this problem.*

- *Machine assignment: The system can be configured to only allow certain products to run on certain machines. So in our example, products containing peanuts might only be assigned to machine A while non-peanut products only run on machine B.*
- *Avoid bad sequences: If products with different allergens must run on the same machine, then the system can be configured to disallow certain sequences. For example, products with no peanuts could be prevented from directly following products with peanuts (unless a major cleanout were scheduled between the runs). So all peanut-free products might run at the beginning of the week, saving peanut-containing products for the end of the week. In the event that a 'bad sequence' was absolutely necessary, the system can display a visual Flag (a diamond shaped block) plus schedule adequate time for a major cleanout to take place.*
- *Cross-machine contamination: In situations where airborne particles can result in cross-contamination between machines, it's necessary to prevent noncompatible products from running together on machines in the same vicinity."* — Using PlanetTogether for Effective Scheduling in the Process Industry

In the previous paint example, color was the driving factor of the changeover time and cost. With other products, different factors cause different sequences that result in lower costs and time taken for changeovers. It must be recognized that there are trade-offs to emphasizing changeovers and one should not take only changeovers into account when making production sequence decisions. I concede this point, even knowing that many others who have different incentives do not concede the counterpoint. A philosophy that focuses only on minimizing inventories will impose higher costs on manufacturing. The question of how much higher should be evaluated. In fact, let us take the example that we have used thus far to see how such an analysis would work. Here are the assumptions:

1. Each production run is for a week.
2. There is zero lead-time from the end of the production run until the product is received by the customer (this could be possible, if the customer picked up the paint with their trucks from our warehouse).

3. The costs of maintaining inventory per week will be the following formula: The Number of Gallons Per Week x $14 per gallon (the cost to produce a gallon) x (1/52)(one week) x 20% (20% being the carrying cost formula per week)

4. Customers will accept late shipment without penalty, but after one week, they expect a two percent discount.

Inventory Carrying Cost / Changeover Trade Off

	Demand for Week 1	Demand for Week 2	Demand for Week 3	Gallons Produced Per Week	Inventory Carrying Cost
Ex. 1	Yellow	Dark Blue	White	3,000.00	$ 484.62
	Costs Saved by Keeping with Lowest Changeover Cost Schedule				$ 432.05
Ex. 2	Yellow	Dark Blue	White	6,000.00	$ 969.23
	Costs Saved by Keeping with Lowest Changeover Cost Schedule				$ (52.56)

If we observe the example above, there is a discrepancy between the demand "sequence" and the optimal production sequence. (We will look only at costs, because trying to look at both costs and duration would make the example more complicated than it needs to be.)

The optimal production sequence for changeovers is White-Yellow-Dark Blue. However, the schedule that minimizes the inventory is one that is synchronized with demand, which is Yellow-Dark Blue-White. If we decide to stick to the optimal changeover production schedule, the Yellow batch must be held for one week and therefore will be one week late (but as stated, customers will accept their product late by one week without penalty), the Dark Blue for one week, and the White for two weeks. This means that there is a three-week period of wait time. The wait time is the natural outcome of the combination of a production schedule that is based upon changeover minimization and the demand per paint color per week. Because of this discrepancy, there is a conflict between the production schedule that would minimize inventory carrying cost and the production schedule that would minimize changeover costs. Therefore this intelligently run company has a policy of basing the production schedule upon the lowest total cost. (In real life, total cost would include more than just inventory carrying cost and the changeover cost, but again, for simplicity, I am only showing these two costs.) In the first example with a production level of 3,000 gallons per week, it is more economical to keep to the optimal changeover. However, if the production level were to be 6,000 gallons per week, it would begin to make sense to change the production sequence.

A topic of conversation for the chemical industry is "product wheel" scheduling, where it is advantageous to follow a predefined production sequence for producing numerous different polymer blends. The production sequence can be complicated with the existence of "side wheels," **which contain products that may or may not all be needed,** but will have an opportunity to be produced when the main product wheel gets to the point in the cycle where the side wheel is encountered. Breaking into the wheel is possible, but highly undesirable.[74]

Trade-offs in production planning can be quantified using a software application. One of the benefits of software is that the values can be added to the application and any hypothesis can be tested. In order to do this testing, an application should have strong simulation capabilities—and this means a thorough evaluation of the software in this regard because many more vendors say they have effective simulation capabilities than actually do and there are often major differences in simulation between applications in the same software category. SCM Focus has a site dedicated to simulation, and believe that a strong weight should be given to simulation during supply and production planning software selection. However, simulation is a universal requirement for all the manufacturing environments and not just specific to process industry manufacturing.

http://www.scmfocus.com/supplychainsimulation/

Before we move on from this topic to how changeovers can be configured in PlanetTogether, it is important to note that changeovers also relate to work in process. In the example that was just provided, the problem was simple in that the trade-off was between inventory carrying cost and the changeover cost. However, what if a product must be processed after a certain period of time or it will degrade

[74] Dave Biddle, Senior Application Consultant, PlanetTogether

or be lost entirely? This is the spoilage issue. It turns out that spoilage can be modeled in PlanetTogether as well. The ability to perform multi-factor decision-making is, of course, an advantage of good production planning and scheduling software. Here is how PlanetTogether accounts for the spoilage issue.

> *"Food spoilage typically results from either product cross-contamination or from products being excessively delayed in the production process or in storage. PlanetTogether can reduce the risk of spoilage due to production delays using the 'Maximum Delay' value in the routing. A maximum duration can be specified between any two operations and if this delay is violated during scheduling, a visual Flag is shown on the Gantt to warn the planner. Pointing to the Flag shows a popup with details explaining that the product is being delayed beyond its Max Delay value. In addition, preferences can be given to material that has been waiting for longer periods of time, thus reducing the chance of this Max Delay being violated at all. Spoilage can also be reduced by producing finished goods and intermediates as close to their need date as possible. This reduces the amount of time the product will sit in inventory, hence lengthening its remaining shelf life. Production orders can have limits placed on them to prevent them from scheduling to start before a Release Date. In addition, each production resource can have a Maximum Head Start duration set to prevent it from running any products more than x days/hours before necessary. Lastly, production can be allowed to schedule early, and simply have a warning 'early' color displayed to alert the planner to the condition for manual control."* — Using PlanetTogether for Effective Scheduling in the Process Industry

Changeover Optimization in PlanetTogether

What I laid out previously is basic or standard changeover functionality. Most production planning and scheduling software has the ability to account for the setup using this input data. However, in addition to this functionality, PlanetTogether has a much more nuanced changeover-planning functionality that is expressed in two other methods of accounting for changeovers. These four different methods are the following:

- Simple Setup
- Product Attributes
- Product Attribute Number Range
- Adjusted for the Specific Resource

What this means is that while most production planning vendors provide changeover planning that can be compared to a butcher knife, PlanetTogether offers more of a scalpel.

Changeover Planning with the Setup Table

This is identical to my previous spreadsheet changeover example. It consists of a set-up code, along with a price and duration. This is the most basic set-up functionality, and all production planning and scheduling software should include this level of functionality. Secondly, while many vendors have this functionality, I have experience with at least one application (PP/DS) where this functionality is difficult to use. Having such elementary, yet difficult to use functionality is a major red flag regarding the overall development quality. Therefore, I recommend that companies actually test using this functionality; sit with the vendor application consultant, configure the functionality, and see it being accounted for in the planning results. Now let's move on to how this set-up table is managed in PlanetTogether.

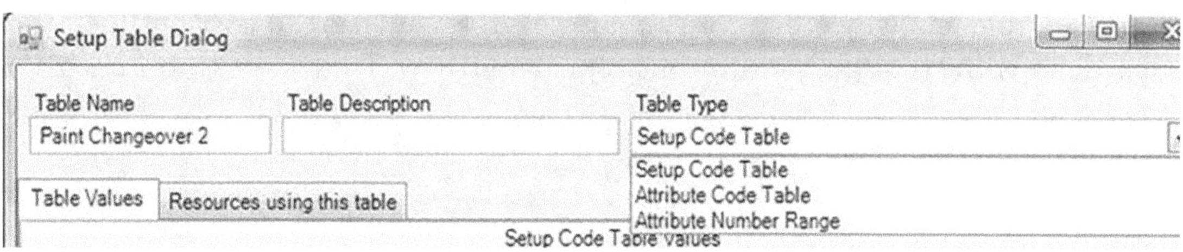

Set-up Code Table: The scheduling engine matches the set-up code values of each operation to the values in this table to calculate the operation set-up time.

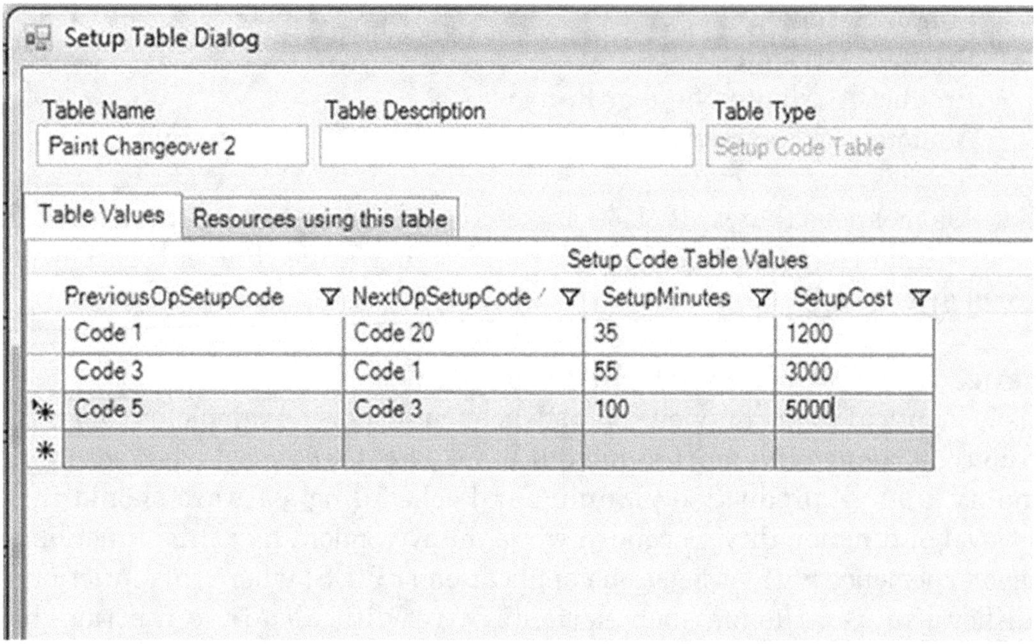

This is the simplest of the three methods in PlanetTogether and it does not have any further detail aside from one code to the next. It is interesting how PlanetTogether deals with the time and costs entered in the Setup Table. PlanetTogether's main optimizer engine is a duration optimizer. That means it attempts to choose the solution that meets demands yet shortens the total production time. There are numerous reasons why I believe this approach is superior to cost optimization, but I don't want to digress into that topic right now. I have written articles on this very topic for those who want to learn more, and to find out information that is very rarely published.[75]

> http://www.scmfocus.com/productionplanningandscheduling/2011/01/20/cost-vs-duration-based-optimization-for-scheduling/

Therefore, the primary value from the Set-up Table that the PlanetTogether optimizer will be trying to minimize is the SetupinMinutes field. PlanetTogether can apply optimization adjustments, which can adjust the optimization on a wide variety of criteria, one being cost. If this cost adjustment were used, then there would also be a weighting for the SetupCosts. However, the determination of whether or not a cost should have its weight increased would have to be made for all the costs entered in the application, so it is not simply a "one off" decision.

Changeover Planning Using Product Attributes

The second way of performing changeover management is with product Attributes. This is where the changeover functionality in PlanetTogether gets quite interesting. For illustration purposes I have added Color, Density, Viscosity and Oil Content as being the most important "attributes" in the setup.

---

[75] While it is often presented as such, cost-based optimization is no longer a best practice for production scheduling (nor is it a best practice for supply planning). Attempting to cost optimize all of the areas of the supply chain was a first attempt to bring optimization technology to supply chain management. However, there are many forms of optimization, and one can set anything as the objective function. In the case of production scheduling, duration-based optimization is far more practical, implementable and realistic for most production scheduling environments. This is because durations are much less open to debate than costs, and the actual goal or "objective function" is to minimize time, not costs.

> http://www.scmfocus.com/supplyplanning/2011/07/10/customizing-the-optimization-per-supply-chain-domain/

As with the concept of PP/DS, values entered in the Set-up Table are then used by the optimizer in the scheduling application to determine an optimal sequence, partially based upon the differential set-up times. However, the optimizer in PP/DS and PlanetTogether work differently. The optimizer in PP/DS works on the basis of costs, while PlanetTogether works on the basis of duration or time. For this reason, costs must be entered into the PP/DS optimizer, while set-up time/duration is entered into PlanetTogether.

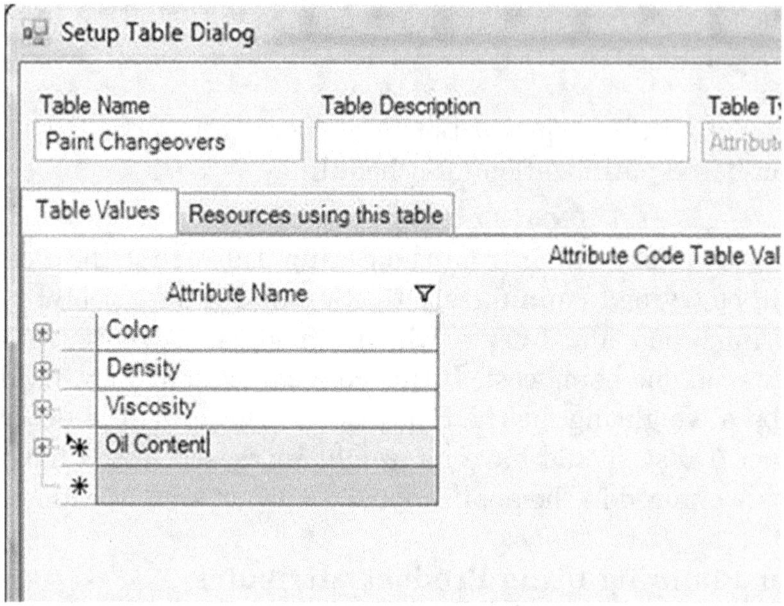

Each one of these attributes has a set-up table associated with it. This allows multiple attributes to be included in the optimization.

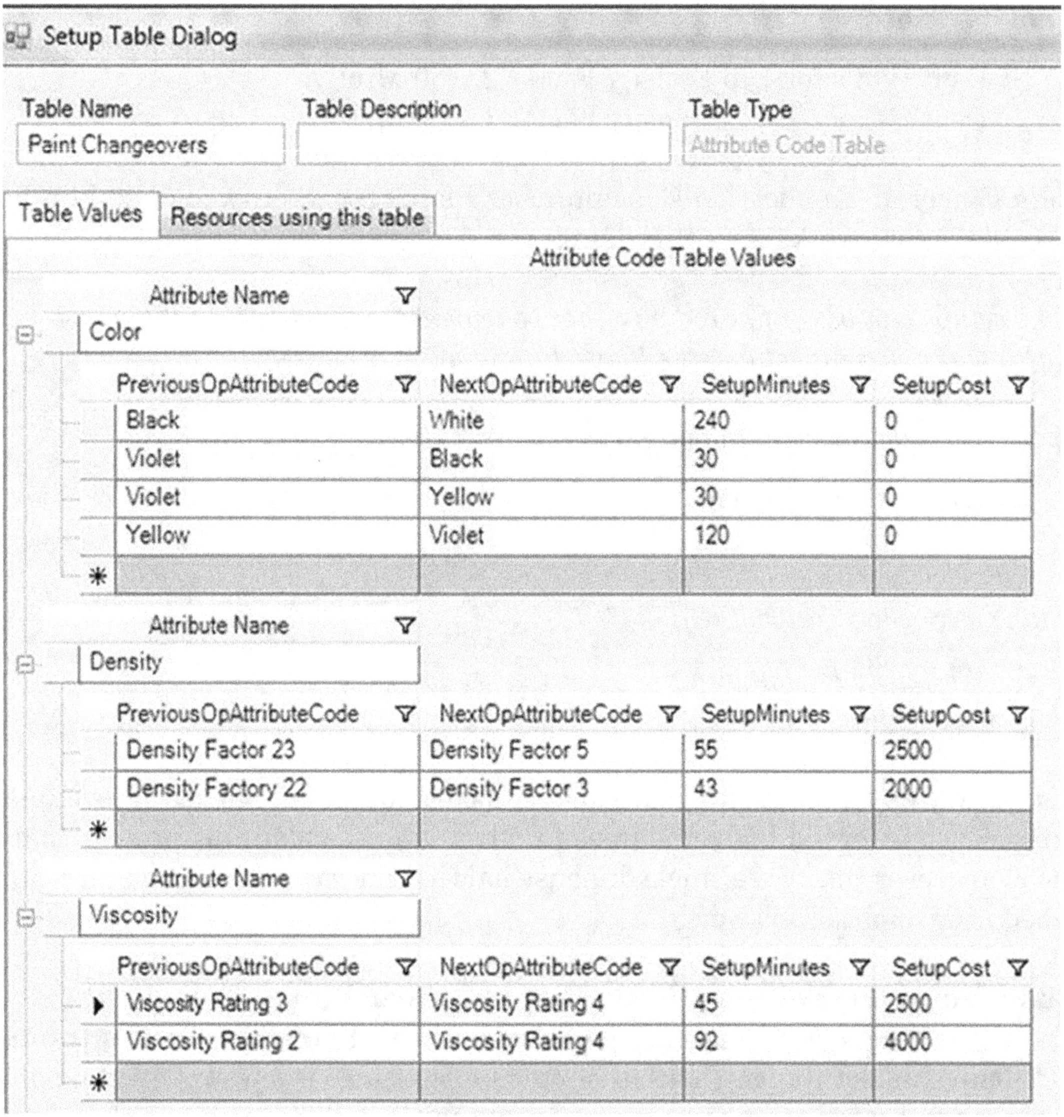

Attribute Name: Color

PreviousOpAttributeCode	NextOpAttributeCode	SetupMinutes	SetupCost
Black	White	240	0
Violet	Black	30	0
Violet	Yellow	30	0
Yellow	Violet	120	0

Attribute Name: Density

PreviousOpAttributeCode	NextOpAttributeCode	SetupMinutes	SetupCost
Density Factor 23	Density Factor 5	55	2500
Density Factory 22	Density Factor 3	43	2000

Attribute Name: Viscosity

PreviousOpAttributeCode	NextOpAttributeCode	SetupMinutes	SetupCost
Viscosity Rating 3	Viscosity Rating 4	45	2500
Viscosity Rating 2	Viscosity Rating 4	92	4000

Let's say that the following applies:

Paint Color Attribute

- White:Blue:100 Min
- Blue:White:200 Min

Density Factor

- Density Factor 1 to Density Factor 2 – 50 Min
- Density Factor 2 to Density Factor 1 – 30 Min

John Cameron, Application Consultant at PlanetTogether, would explain how PlanetTogether works with this data input.

"The setups can be configured as either concurrent or consecutive. When you have multiple concurrent setups, the longest (time-wise) is chosen.

In the example above, if they were concurrent, then:

- *White:Blue = 100 min*
- *Blue:White = 200 min*

If they were consecutive, then:

- *White:Blue = 150 min*
- *Blue:White = 230 min"*

Basically, this allows multiple attributes to be accounted for in changeover planning, which inserts a higher degree of realism into the planning process than if the changeover time were single dimensional (how most production planning and scheduling applications work).

Changeover Planning with Attribute Number Ranges

The third approach to managing changeovers in PlanetTogether is through Attribute Number Ranges. The attribute number range is a way of dividing products that fall into specific ranges, as shown in the following screen shot. Using this functionality we can specify set-up times based upon a range of values.

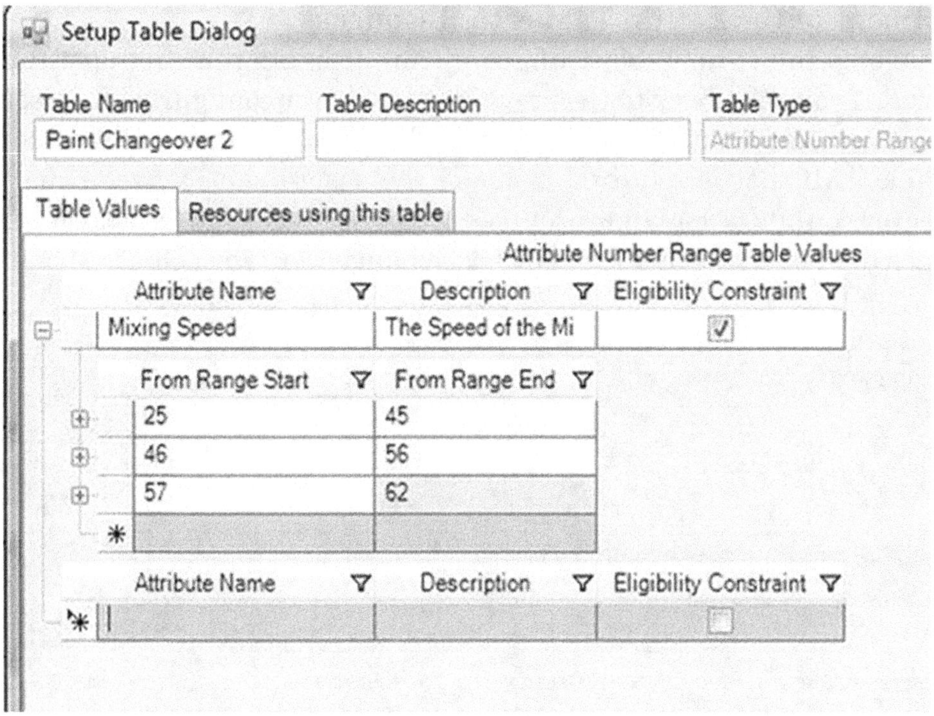

In this case the changeover is based upon the number range of the mixing speed.

Instead of basing the changeover on attributes of the product (such as its color), this functionality bases the changeover on the actual range of the ***production operation process***. For example, in the screen shot above, products are classified in terms of their changeover time, which is based upon mixing speed. However, the range could be anything—oven temperature is another example of an attribute number range. This functionality is particularly useful when the changeover time correlates with the range of the process. For instance, let's say an oven must be turned up very high for baking one product. Is it better for the next product to be one that requires a lower temperature or a higher temperature? The changeover will take longer if the oven must cool off than if the changeover is to a product that is baked at a higher oven temperature. PlanetTogether has learned from various clients that sometimes the product—and other times the process—drive the changeover time.

Assignment of the Set-up Tables to Resources

The final step in Set-up Table configuration is assigning the Set-up Table to the resources. If you refer back to the previous changeover configuration screen shots, you will see a tab next the table values. This tab is named "Resources using this Table." All that is required to make the assignment between the specific changeover configuration and resources is to select that tab, and then select the resources that should be managed by each changeover type. This is shown below:

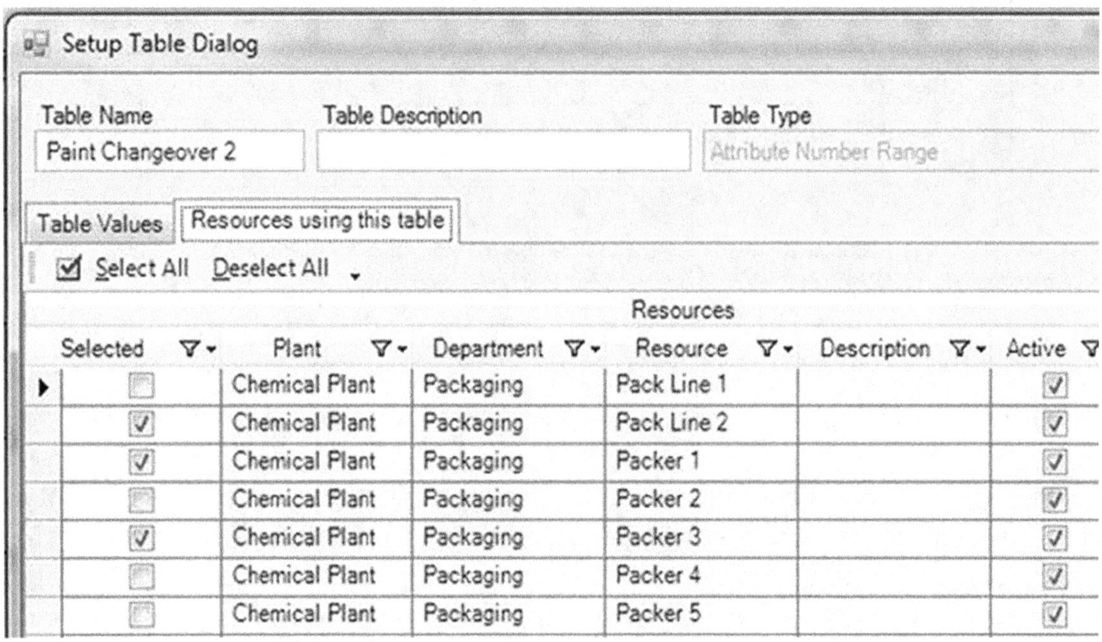

The resources that apply to this changeover configuration can be assigned simply by selecting the check boxes under the "Selected" column. All of the resources that have been set up in the application are shown; however, only those resources that should use the particular configuration should be selected.

Which Sub-method should be Used?

One or all three of the methods for planning changeovers can be used at any installation, and the resources can be flexibly moved between the different resources. Most production planning and scheduling applications do not offer more than one method of changeover planning, and once configured, the changeover can be difficult to adjust. PlanetTogether's functionality meets the following important requirements of changeover functionality:

1. Ease of use
2. Ease of maintenance
3. Flexibility in how the configuration is created
4. Flexibility in how the configuration is changed

Up until this point we have discussed the assignment of resources to specific methods of performing changeover planning. However, specific adjustments are sometimes necessary to the resources even within a particular changeover method. Now we will move on to the next interesting changeover functionality.

Adding Resource Specificity to Changeover Planning

PlanetTogether is not limited by the setups per product attribute, but the application can also account for ***particularities of the resource***. Within the resource there is something called the SetupSpan field, shown in the following screen shot.

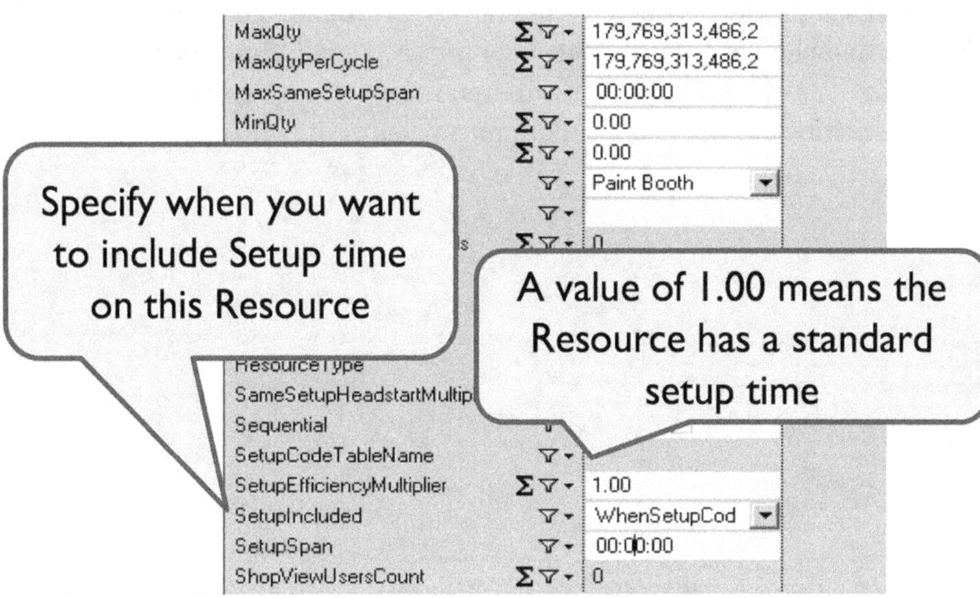

This field is a multiplier for how long this resource takes to changeover versus other resources. For instance, if a particular resource takes fifteen minutes longer than other resources to changeover, by adding this time element to this field the system will account for it when it performs changeover planning.

PlanetTogether actually assigns two fields to the resource that are similar to one another in this regard. The other is SetupEfficiencyMultiplier. This is the same concept, but instead is entered as a percentage. If a particular resource takes 10 percent less time to change over than other resources, the percentage can be added here and again the system will account for this in its resource calculations.

Conclusion

In manufacturing generally it is common to see the terms "setup" and "changeover" used interchangeably. In process industry manufacturing, a big part of the changeover process is the cleanup or cleaning. The main point of cleanup is to both enable the next processing operation to operate smoothly and to prevent the previously produced item from polluting the next product to be produced. Changeovers apply more to batch than continuous production, but they apply to both in that while a continuous process is not actually stopped, the transition can result in finished product, which is outside of the finished good specification and is therefore not usable—at least for sale as a finished good. Some types of continuous production are never changed over—instead they simply run as long as possible in a 24/7 scenario reaping the gains of many years of continuous production.

CHAPTER 9

Constraint Planning and Process Control

All the types of manufacturing environments tend to have decision makers that want to constrain their resources so that only feasible production orders are created by the planning system. Constraining manufacturing resources that do not change frequently is one thing, but a major complexity of process manufacturing is the variability of its constraints. In this chapter we will cover how to effectively model constraints, as well as constraint variability.

Resource Variability in PlanetTogether

PlanetTogether is capable of managing resource variability quite well, which I will demonstrate in a just a few paragraphs. I should say that this ability (or a similar ability) should be requested by buyers of other production planning and scheduling applications, particularly if these applications are intended to be used in the process industry. PlanetTogether has the following to say about how resource variability is accounted for in their application.

> *"Often times, the output rate and expected yield for a single product can vary depending upon the specific piece of equipment used. For example, one mixer may produce 250 gallons*

per batch while another produces 1,000 gallons per batch so the time to produce x gallons varies based on the mixer chosen. ***This is handled by creating a lookup table that specifies the capacity, production rate and expected yield of each product on each machine. In this case, PlanetTogether will dynamically schedule using the appropriate specifications for the given product on a given machine.*** *If you 'drag-and-drop' the production run onto another machine, the time will be updated automatically and it will also be updated if the production run is reassigned during the Optimization process."* — Using PlanetTogether for Effective Scheduling in the Process Industry

Variability is frustrating to model because specific output quantities cannot be depended upon. Part of the job of process manufacturing is to control this variability (this is referred to as "process control") and this is performed outside of the planning system and there are steps designed to reduce the actual variability in the factory.

Process Control

Process control software is used for maintaining control over processes within ranges. A good example of everyday process control is found in automobiles in cruise control, automatic temperature control and anti lock brakes. Previously, cars required manual adjustment to heating and cooling the cabin, however, with process control, many modern cars can now simply set the temperature and the process control unit will keep the temperature at that desired level. Process control is important for the quality of the finished good/output product. For instance, in bread baking (as is covered in Chapter 2: "Continuous Versus Batch Processing"), bread can be baked quickly with high heat—however, a baking process that takes longer and is set cooler will result in a better tasting bread.[76] A baking oven is a simple single process to control—set with something as simple as a thermometer and a timer. Process control also affects the preparation of the input product. If we take the example of bread again, if dough is overproofed, the yeast will run out of food too quickly, and the resulting bread will taste acidic. The objective of process control is to keep the critical production factors in the right range for the

[76] Understanding the Art and Science of Baking

entire production process. We have referred to the use of yeast in various food manufacturing processes for fermentation. However, fermentation is an excellent example of a process that must be tightly controlled and that process control is ideally suited, as the following quotation explains.

> *"Fermentation is the process by which living organisms are cultured or grown to produce a specific product. The products can be as simple as baker's yeast and alcohols, or as complex as therapeutic proteins, antibiotics, enzymes, and genetically engineered materials. Regardless of the end product, fermentation reactions must be carefully monitored to gauge the health of the culture, and to determine when the culture can be harvested. Since these are living organisms, the levels of oxygen and carbon dioxide in the off-gases can be used to follow the progress of fermentation. The ratio of carbon dioxide to oxygen is called the respiratory quotient (RQ) and is the prime indicator of culture health. While these measurements can be made by several techniques, the most accurate RQ determination requires complete knowledge of the flow of gasses used previously into and out of the reactor. By virtue of its ability to measure all of the gases entering and exiting the reactor, the ProLine process mass spectrometer provides a highly refined control signal to maximize culture health and productivity."* — Control of (the) Fermentation Process

Process control is based upon control theory, which is how to control the behavior of dynamic systems—which extends to fields such as engineering, mathematics, psychology, sociology, criminology and financial systems.[77] The four basic functions of control systems are:

1. Measure
2. Compare
3. Compute
4. Correct

[77] Wikipedia

One early application of control theory was the centrifugal governor.[78] Process control is easier in linear systems, but has been applied to non-linear systems as well. Process control is made up of the following data and process:

"Monitoring and Sensing

- *Process*
- *Equipment*
- *Environment*
- *Labor*
- *Material*

Control

- *Machine Control*
- *Regulatory Control*
- *Real Time Quality Control*
- *Advanced Process Control*

Operations

- *Process Sequencing*
- *Machine and Process Instructions*
- *Labor Instructions*
- *Human Machine Interfacing*
- *Safety*
- *Maintenance"*[79]

[78] Wikipedia
[79] Manufacturing Execution Scheduling

Common Uses of Process Control in Manufacturing

Process control is intensively used in continuous process manufacturing with some of the most common users being petroleum refining, paper manufacturing, chemicals and power plants.[80] The more complex the process the more important is process control in obtaining the desired output product. For instance, in petroleum refining the thermal and catalyst cracking process is complex and must be tightly controlled in order to generate the desired output. The feedstock should have been optimized using planning software, but the role of the process control software is to make the production process **actually match** the simulation that was performed during planning. In order to optimize the process control of complex processes like cracking it is necessary to archive data, where again the role of the software data historian comes in.[81] If yield was better for one production shift, the data historian software provides the ability to go back and review what was done correctly so it can be replicated. Once understood the process control software can be calibrated to the specifications that were used during the period of the best yield. These settings can be documented and then serve as the intellectual property of the company.

> *"During plant operations, the optimum conditions can change frequently owing to changes in equipment availability, process disturbances, economic conditions (for example raw material costs and product prices). Consequently, it can be very profitable to recalculate the optimum operating condition on a regular basis."*[82]

[80] Wikipedia

[81] Cracking is the process of breaking complex heavy hydrocarbons into simpler hydrocarbons by breaking the carbon-to-carbon bonds with a combination of high temperatures and pressures and sometimes with the use of a catalyst. This is different than distillation because it actually alters the chemistry of the crude oil rather than simply separating it.

[82] Process Dynamics and Control

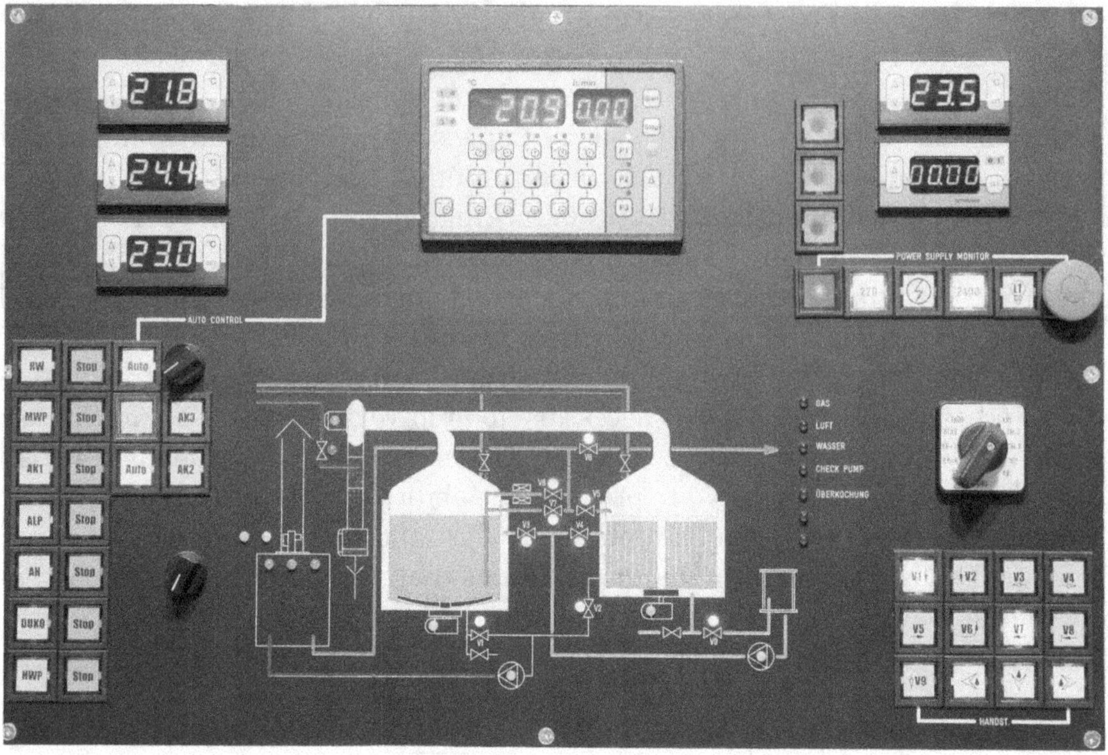

This is a process control panel for a brewery. The vats are shown in the middle of the panel, and lights tell the user if the different parts of the manufacturing process are within specification, if not the lights change color and the panel produces noises to get the attention of the user to make the necessary adjustments to bring the process back into the proper specification.

These types of process control panels have been installed in plants, but in the future they will be increasingly computerized.

A purely software process control panel is far more flexible than a hardware process control panel and allow more individuals to monitor the production process from any distance, and enables the development of more advanced monitoring and alerts.

All of this has the following implications:

1. *Increased Monitoring Input:* Plants can be monitored by more people, and even at a great distance from the plant. This can be quite advantageous in that one can have more eyes on the panel's indicators, but also it can help mitigate specific skilled labor shortages in certain regions, which is an issue faced currently in Asia with respect to process industry plant management and monitoring knowledge.
2. *Mobile Monitoring:* Process control software can be run from mobile devices such as tablets and phones.

3. *Historical Data:* The computerization of process control panels not only provides better and more flexible monitoring of the plant, but also for reviewing the historical data, which due to computerization, can now be archived and mined for insights. This is referred to as product tracking; the AspenTech product Aspen Production Record Manager is one such product. Here is a scenario that explains how this historical operating data can be leveraged.

4. *Integration with Other Plants and Analytics:* With computerization of process control data and data archival, the process control data from multiple plants can be shared in order to improve the processing at any one plant. Data can be reviewed as a report or extracted for statistical analysis. Fully leveraging this data may require a different category of software called data historian.

5. *Integration with Other Applications:* Once transitioned to a software application, the process control function can be integrated into other manufacturing applications described in this book.

The Origins and Logic of Computerized Process Control

Computerized process control was first analyzed in the mid 1950s for the chemical industry, which resulted in a Texaco refinery and a Monsanto chemical plant using mainframe computers to support their process control system.[83] This provides some insight into the importance of process control to certain process industry sub-categories.

An example of how the computerization of process control can change plant management is explained in the following quotation.

> *"Let's assume a process manufacturer is headquartered in the United States with a plant in Asia. Two customers call the corporate customer service group to report contamination issues that are discoloring their final products. The customer service group in the United States has access to process data from the plant historian, but the company does not have any specific product tracking tools. To troubleshoot this issue, the customer service engineer would first try*

[83] Process Dynamics and Control

to determine what vessels, raw materials, and additives were used to produce a product before looking for any identifiable root causes. He would also need to determine which other lots or campaigns used the same equipment, materials, etc. and investigate whether those batches had similar issues. This is product tracking the hard way and can take significant amounts of resource time to fully determine the scope of the problem before getting to the root cause investigation. This approach delays the identification of other customers who could be affected by the issue. In addition, while the investigation is taking place the site is still producing products that potentially have the same contamination issues. If this company had a production management system for marking, tracking, trending, analyzing, and reporting of production and event data...the ***customer service engineer would be able to quickly and easily find specific batches, campaigns, or lots based on process data, equipment used or other production data of interest.*** *[emphasis added] The engineer could then proactively notify other customers who might have the impacted products. He would also immediately be able to focus on identifying the root cause of the contamination to minimize liability risks and potentially reduce or eliminate additional product quality issues. In this example, the goal of the customer service engineer is to quickly identify the root cause of the contamination. Without the proper tools, the process of getting to the root cause can involve a significant amount of analysis work to determine basic information associated with the production campaign."* — Enabling Process Manufacturers to Leverage Resource Expertise Globally

The process control application is where the process instructions are kept. Process instructions are important in process industry manufacturing and this differs from discrete manufacturing in several important ways and it relates to where the "intellectual property" of the finished good is actually maintained. This is explained in the following quotation.

"The 'how' of assembly in discrete manufacturing is easily reverse-engineered, yet the components may be patented. In contrast, the 'how'

of processing is the focus of patents. Process instructions are often the key to higher yields and better quality using the same material inputs as a competitor's. The secret might be a specific parameter held constant during the processing, or a specific technique in heating a mix. Often the secret is in the interaction of process steps and ingredients introduced sequentially into the process. To support consistent yields and quality, however, these process instructions must be communicated clearly to the operators. Because a BOM-based system assumes that the materials listing is separate from the production steps (they are held in a separate file called the routing file), there are risks in correctly communicating nuances of processing to operators. In a chemical ERP modeling system, the process stages clearly show the sequence of each processing step and which inputs are introduced at each stage. This type of architecture allows modeling expertise—the manufacturing process—and enables improvements to product quality and overall competitiveness."
— Process Industry ERP Requirements

Sources of Variability

Even with process control software, variability still exists. This variability is such a constant issue with process industry manufacturing that it can be difficult for people from other manufacturing environments to appreciate how much this variability affects the production process. There are several different sources of the variability.

1. *Variability of the Feedstock*: One source is the input material, called the feedstock.[84] The qualities of the materials vary greatly over time and place. Process industry manufacturing companies need to be able to work through and around these variations to make a finished product. Discrete industries are often the receivers of the output of the process industries and therefore are starting with materials that have **already** been adjusted to a desired specification for their use.[85]

[84] As mentioned earlier in the book, feedstock is the raw or input material and is often a natural material: apples, corn, crude oil, trees, iron ore.
[85] Dave Biddle, Senior Application Consultant, PlanetTogether

2. *Output Yield Variability*: While discrete and repetitive manufacturing has scrap, the output variability is nothing like the yield variability of process industry manufacturing. For example, in planning for a milk processing company, the amount that it has rained, the time of year, along with a host of other factors changes the percentage of the output product (milk, whey, curd, pharmaceutical fillers) that could be produced for a given quantity of input product. This means that if, for example, the production plan called for 4,000 pounds of whey to be produced at a plant for a particular day, the plant may or may not have the raw material to produce the quantity specified.[86] Instead the yield could be higher in another product.
3. *Production Variability*: This is ostensibly the variability over which the process control software can directly influence. The point is to minimize this variability for a certain production specification. A precise and tuned process control system can allow fine control, which both maintains constantly for a particular specification, but can also change the controls for different specifications.
4. *Supply and Demand Variability*: This is the age-old issue of unpredictable changes to supply and demand.

Even after process control has been fully utilized, variability must normally be managed **through carrying higher stock levels**. In discrete manufacturing, there is supply variability and demand variability (demand and supply variability are the inputs to the calculation of dynamic safety stock as is explained in the following formula). Supply variability is often thought of in terms of supplier lead-time variability, but of course there is more to it even in discrete manufacturing. The dynamic safety stock calculation is based upon accounting for both demand and supply variability.

[86] There were a number of problems with variability that I was expected to model with SAP PP/DS. Another was the unplanned removal of resources, such as tanks for maintenance, and for which there was no maintenance schedule. However, in process industry manufacturing, the supply variability must be expanded to include manufacturing variability.

Safety Stock = {Z*SQRT(Avg. Lead Time*Standard Deviation of Demand^2 + Avg. Demand^2*Standard Deviation of Lead Time^2}[87]

With discrete and continuous manufacturing, production output variability is typically minimal, so generally this formula will estimate the appropriate safety stock level for these environments. However, with process manufacturing, a third variability—yield output—**which is actually a subcategory of supply variability**, must be accounted for as well.

Resource Variability

Another type of variability is the variability of production per resource. Quite an important measurement of any production planning and scheduling system is how well resource variability can be modeled. Limitations on this capability have been a major factor in reducing the success of many a process industry production planning and scheduling implementation.

A Resource's Capabilities

Resources are the entity that actually performs work in a plant. In PlanetTogether's Galaxy APS, Capabilities are the work that resources perform. For instance an example of a resource would be a loom, and the capability of a loom would be weaving. In Galaxy APS, **each resource can have multiple capabilities, and each capability can be linked to multiple resources**. The resource capabilities can be changed easily by selecting the resource from the Resource Tab in the Resource Configurator screen, and then switching to the Capabilities Tab and selecting which capabilities to activate for that resource. This is shown in the following screen shot.

[87] The vast majority of companies do not use dynamic safety stock calculation, and I am not recommending it here, but the formula is the basis for adjusting safety stock even if it is most often actually set with a combination of a static value that serves as the minimum and a days of supply serving as the maximum. You can read more on dynamic safety stock in this article: http://www.scmfocus.com/sapplanning/2009/05/11/safety-stock-formula/

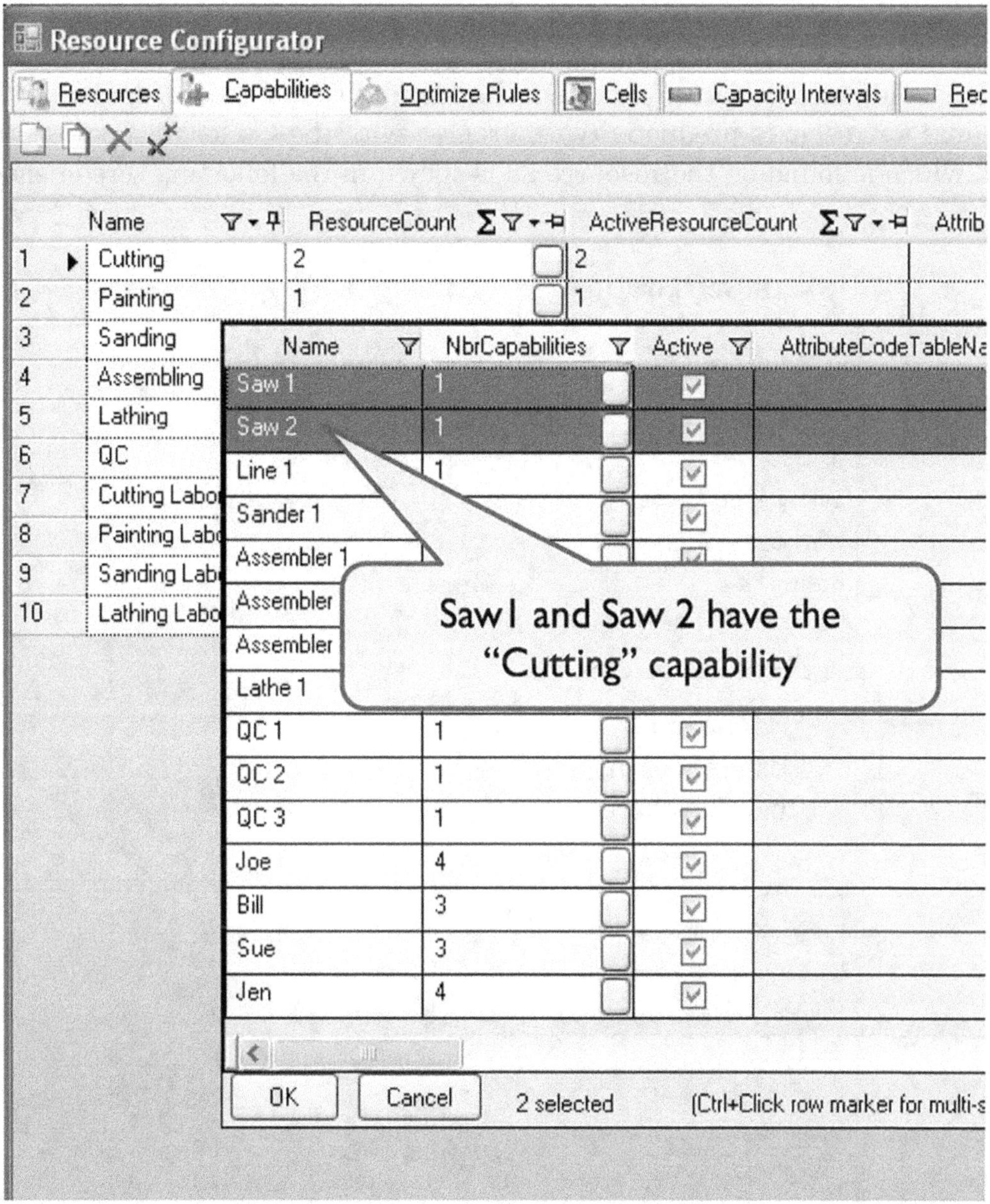

Once configured, these capabilities are respected by the PlanetTogether optimizer during the planning runs as well as when the planners attempt to move jobs using the Gantt chart interface. Galaxy APS will only allow work to be moved in the interface to eligible resources.

Individual Resource Adjustment

While resource capabilities can be assigned flexibly, Galaxy APS also offers the ability to adjust the output of the specific capability per resource. Galaxy APS accounts for resource productivity differences with the CycleEfficiencyMultipler field, which is found on the resource as is shown in the following screen shot.

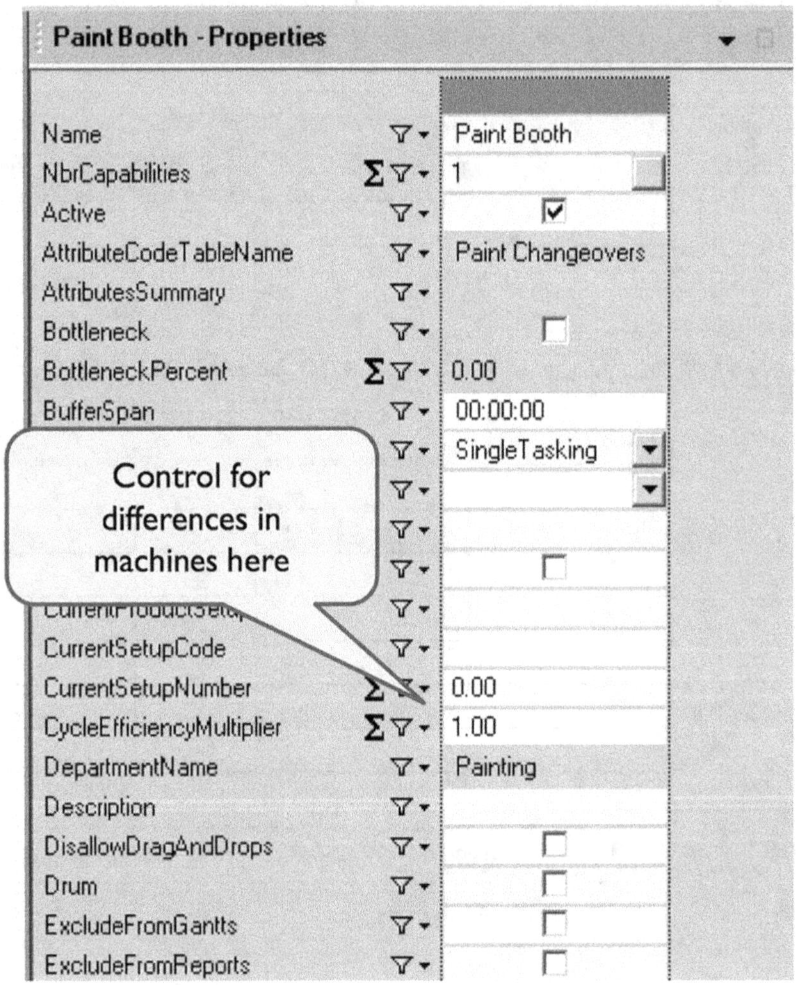

The CycleEfficiencyMultiplier field adjusts the Capabilities, which are assigned to any number of resources.

Min/Max Limits on Resources

Most people who work in production planning are familiar with how capacities are set on a resource in whatever application they use. However, it can also be important to set minimum limits, so that a resource is not used unless it can run a minimum amount of product through for a planned order. Minimums and maximums can be defined in Galaxy APS for the entire manufacturing run, or per cycle, which means for the period of time that a product is in the manufacturing process.[88]

Production Cycle Time

Apple Processing	*1 Apple*	*500 Apples*
	In Seconds	
Peeler	20	10,000
Slicer	30	15,000
Packaging	45	22,500
	Cycle Time	Production Time
Total in Minutes	1.58	791.67

In this example, while the overall production processing time is close to 791 minutes (the time to process 500 apples), the manufacturing cycle time (the time to process a single apple) is 1.58 seconds. Therefore, setting a minimum or maximum on a resource at either the cycle or the overall processing time provides a finer level of control in the application.

[88] *"The period during which the objects of labor (raw products and materials) remain in the production process, from the beginning of manufacturing through the output of a finished product. In addition to the working time, the production cycle includes interruptions in production owing to physical, chemical, and biological (natural) processes (for example, the period required for tanning leather); the character of the objects of labor; or the technology and organization of production."* — http://encyclopedia2.thefreedictionary.com/Production+Cycle

Non-Linear Resources

Process industry resources will often be calculated on the basis of non-linearity. That is, for a 1,000-pound mixer, if 500 pounds of input product is added, the mixer will not complete the operation twice as fast. This non-linear consumption applies to both the planning system as well as to costing, which means to the ERP system. In this same example there may not only be the obvious maximum (1,000 pounds) but a minimum as well—as was just discussed. However, in this scenario the mixer may have a minimum, which is related to the operation of the mixer. So if the minimum is 200 pounds, and the demand is 1,150 pounds in a given planning bucket, and the desire is to meet the demand in the planning bucket, the operation must be split so that both mixing operations are over the 200 pounds minimum.

Resource Variability in SAP APO

Process industry resources have different costs as well as different capacities. The standard way of accounting for variable production per resources in a cost optimizer (such as the one inside SAP PP/DS) is to assign a cost to the master data container that contains the resource. (The container is called a PPM or PDS. These master data objects contain the BOM, the routing and the resource.)[89]

Different versions of the resource allow PP/DS and SNP (the production planning and supply planning modules in APO) to work off of modified resource capacity by time of day, by week, or by month. This can help in making the resources "flex" through time to match the reality of the resource. Different variations of capacity can be stored in the system and can be used for different planning runs. When the system cannot be made to accommodate variability, this means that planners must override the system more frequently, undermining the entire credibility of the system. Much of the poor reputation that planning applications have received in process industry manufacturing planning has been exactly because of applications provided by SAP where the factors that make an application work properly for process industries were simply added as an afterthought.

[89] If you are interested in reading about these master data structures, you will find an explanatory write-up at this link: http://www.scmfocus.com/sapplanning/2009/04/24/pds-vs-ppm-and-implications-for-bom-and-plm-management/

This high resource maintenance overhead that shows up in SAP APO affects SAP ERP as well, so companies that used the production planning module in ERP (PP) have faced the same issues for many years now. Therefore, companies that want to use resource variability in SAP must be willing to allocate a significant number of hours above what other planning applications require in order to obtain the necessary resource flexibility. However, in reality, companies are simply not willing to allocate this type of manpower to this function. Planning departments are run with the low staffing level, and this is true of the largest companies—they would simply prefer to put their money into things. The type of money necessary to support such high maintenance solutions is simply unlikely to be available to planning.

Resource Variability in Petroleum Refining

In refineries, in addition to the variability in the feedstock, other common areas where the quality of the output varies is in the equipment performance over time due to furnace fouling, catalyst degradation, ambient temperatures, and who is running the process. This is explained in the following quotation.

> *"With simple separation processes such as distillation, refineries are at the mercy of the type of crude they run as they cannot change the yields of various finished products. For example, if a crude oil contains twenty percent gasoline boiling range molecules, a refinery will only be able to distill twenty percent of the crude into gasoline using an atmospheric distillation unit or vacuum distillation unit.*
>
> *Gasoline and the group of products known as middle distillates (jet fuel, kerosene, diesel and heating oil) are the finished products with the highest demand and the largest profit margin. There is, therefore, an incentive to maximize the production of just these two product groups by converting as many hydrocarbon molecules as possible contained in crude oil into gasoline and middle distillates."* — Oil 101

AspenTech and Petroleum Planning

AspenTech is a leader in a variety of software products for the oil and gas industry.[90] The following screen shots give a good indication as to why they are so successful in this industry sub-category.

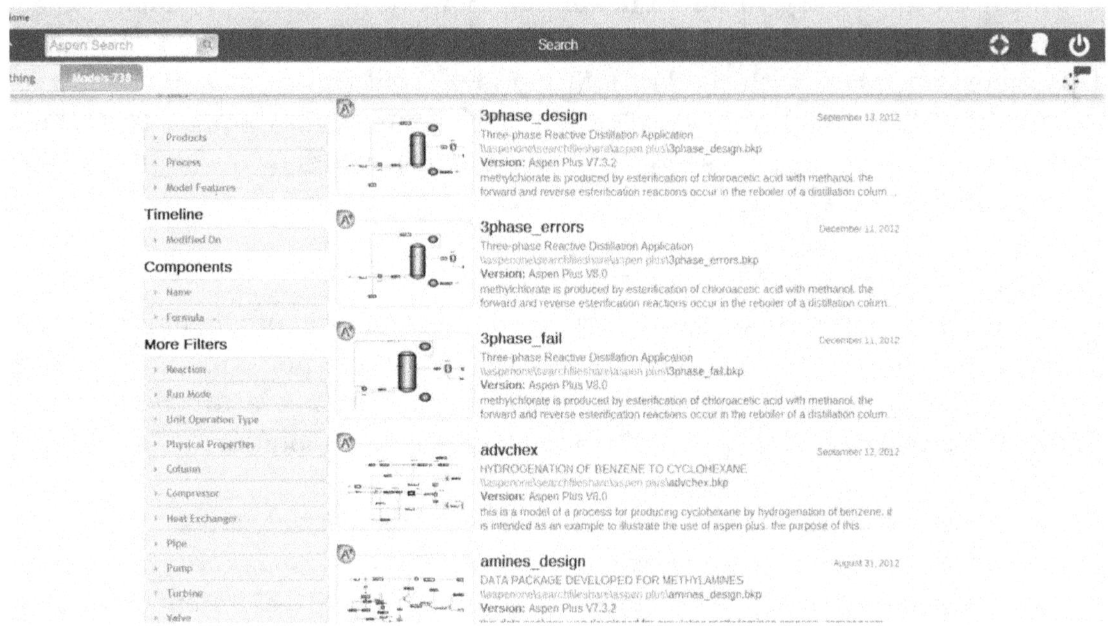

AspenTech has a number of prebuilt model templates, which can be used as a starting point to model various process industry environments. The specificity of the models to be selected is impressive. If we take the first, it is producing methylchorate through a process of esterification of chloroacetic acid with methanol. Because the model is so specific to this process, the settings that are included means that there is very little actual configuration required when compared to other applications. Most applications simply provide software that can be configured and there are very few templates of this type used in any production planning and scheduling application—in fact, there are very few templates of this type used in supply chain planning applications overall. Instead, the predominant

[90] AspenTech is greatly known for oil and gas where they are the undisputed leader for planning solutions—however they also serve markets such as engineering, chemicals, polymers, pharmaceuticals, consumer products and minerals, metals and mining.

approach is to have the model developed from scratch during the implementation. As soon as a model is selected, the specifics of that model are brought up as shown below:

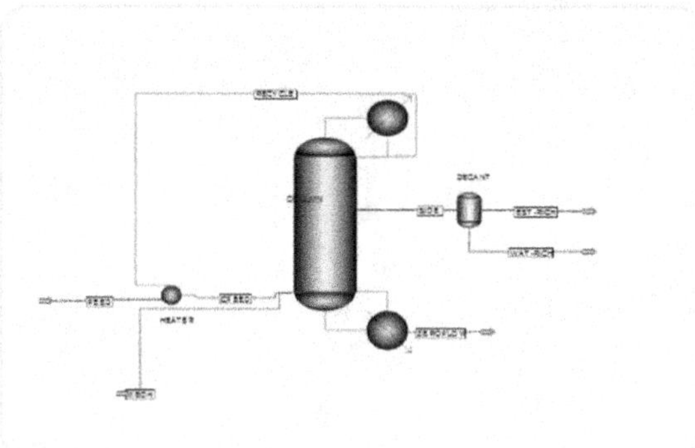

3phase_design

Three-phase Reactive Distillation Application

\\aspenone\searchfileshare\aspen plus\3phase_design.bkp

Product Version: Aspen Plus V7.3.2

Methylchlorate is produced by esterification of chloroacetic acid with methanol. The forward and reverse esterification reactions occur in the reboiler of a distillation column. The column is used to separate the ester product from the reactants. Because ester and water are partially immiscible, two liquid phases occur in the middle section of the column. This application illustrates many unique features of Aspen Plus: 1) Reactive distillation 2) Three-phase distillation 3) Column with total boilup (i.e. zero liquid bottoms flow rate) 4) Internal column specification with feed flow manipulation This application also uses the decanter model.

First Published By: System Accounton 4/11/2013 at 09:16 AM

Last Modified By: on 9/13/2012 at 08:10 AM

Physical Properties: NRTL-RK

Components: METHANOL, CHLOROACETIC-ACID, METHYL-CHLOROACETATE, WATER

Model Features: Reaction Kinetics; Steady-State Simulation

Run Type: Flowsheet

Physical Property Databanks: AQUEOUS; INORGANIC; ASPENPCD; PURE26

Stream Class: CONVEN

User Subroutines:

Linker Options:

This is a model for a three-phase reactive distillation application. This model is a very specific configuration, which is designed for this manufacturing process. This is one of the real strengths of AspenTech—that for oil and gas industries it has such specific standards that can be applied and can control the model.

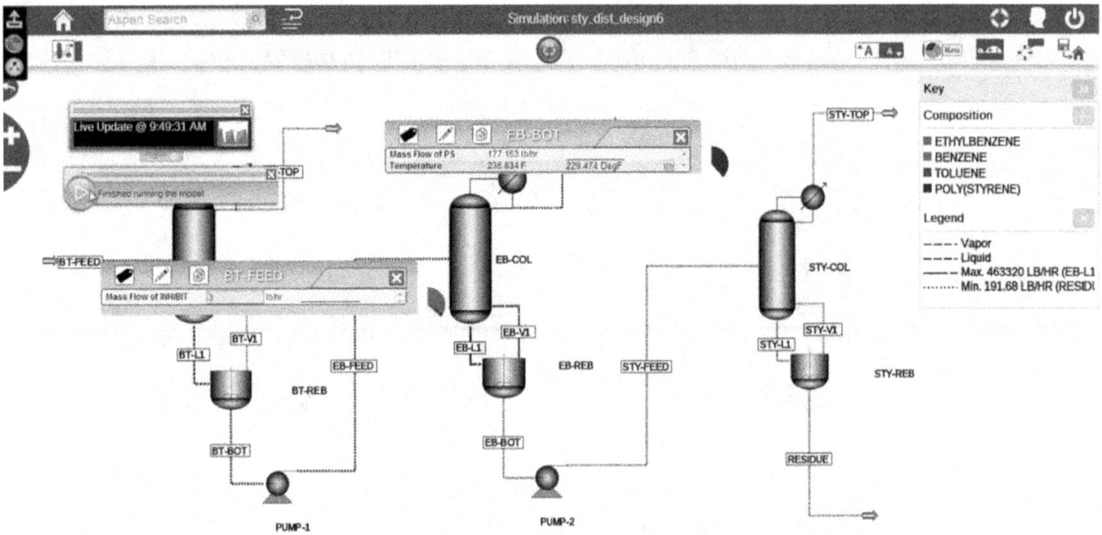

The AspenTech user interface lays out the model that is selected.

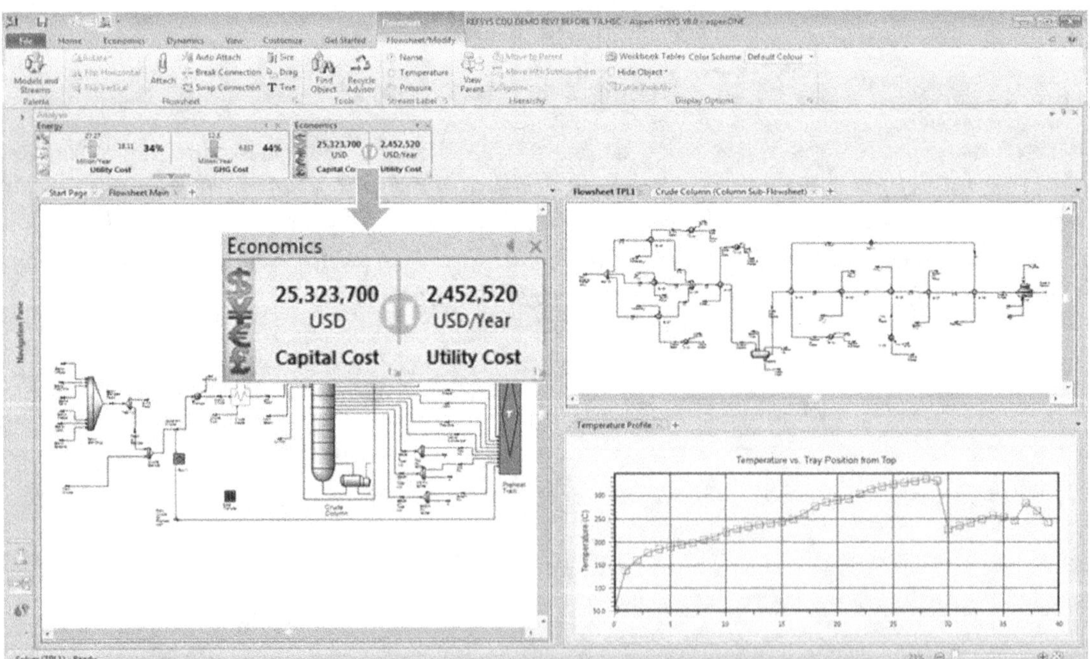

AspenTech can estimate a cost of a particular simulation or simulated planning run. New simulations can be found that result in lower costs (along with desirable finished product). As shown in the previous screen shot, the costs are estimated to be $25,323,700 along with a utility cost. The utility cost is very important because process industry sub-categories such as chemicals, petroleum refining and foods consume energy, water and steam—and their consumption is a major contributor to the cost of production. However, using AspenTech's modeling capabilities, the planner can make changes to settings, which can still bring the finished good in on its specification but will possibly reduce its cost.

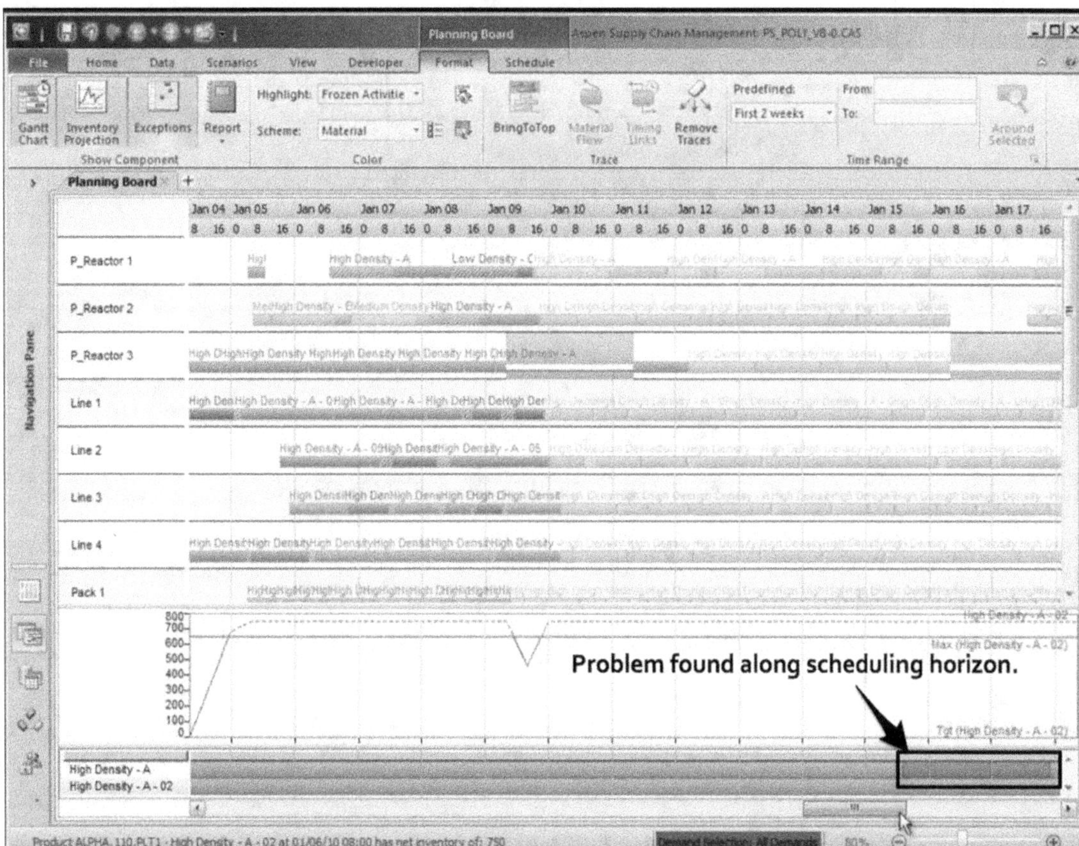

The previous screen shots focus on the modeling side of AspenTech, however this screen shot shows the planning board of the Aspen Supply Chain Management product. This allows the planner to find problems, such as the one above that shows an inventory shortage.

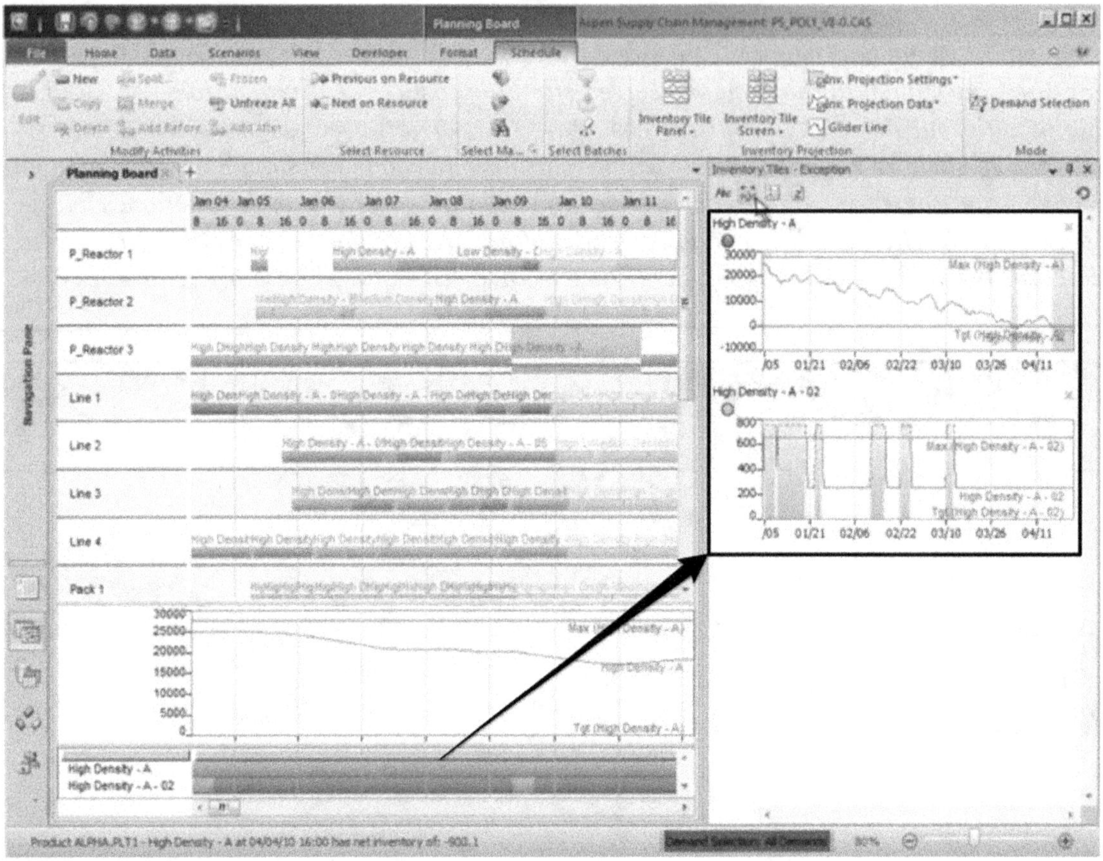

Here, the panel to the right opens showing the inventory for the two products, High Denisty – A, and High Density – A – 02.

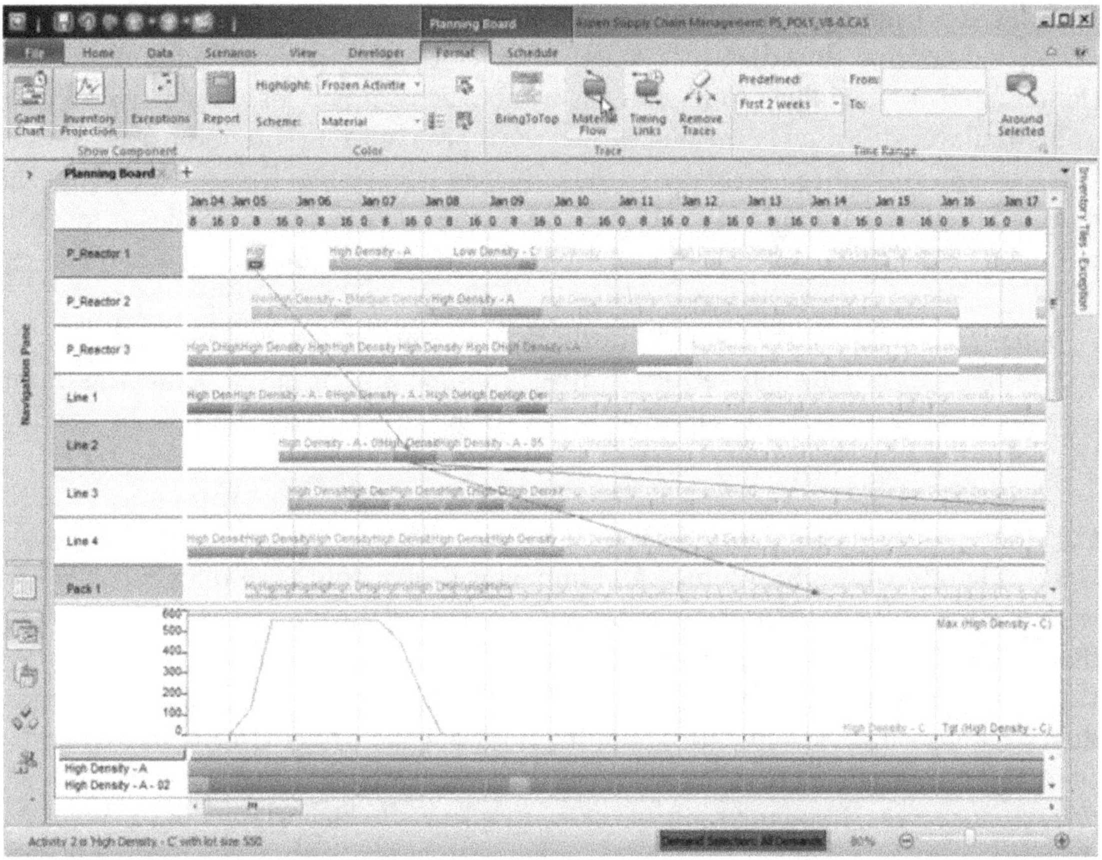

Before inserting a planned order to alleviate this planned inventory shortage, the planner can hit the Material Flow button, which will bring up the product trace functionality, and in this case will tell the planner that the product can be made later without delaying downstream production.

Petroleum refining is a highly complex area in which to plan (depending upon the products being produced). In a typical refinery there can be as many 500 equations and 2000 variables to model.[91] It is, of course, much less expensive to plan and simulate refining operations rather than preforming trial and error. In fact, using complex planning software is the only way to optimize the feedstock and the planned production. Planning with software, which can account for these variables means more of the desired primary product, and less co-product and by-product.

AspenTech actually combines planning with planned process control in order to create not only a planned set of requirements for procurement or for upstream manufacturing, but also a planned set of resource settings that can be used in the actual production run.

Capacity Intervals in PlanetTogether

Ease of master data maintenance is important for resource management; along with the recipe and the routing, resources are one of the most commonly changed master data elements in production planning. This is one part of the application that is often overlooked during the software selection process. While many production planning and scheduling applications set the ***operating times directly into the resource***, PlanetTogether's Galaxy APS ***splits the resource from its operating times***, allowing multiple operating times to be assigned to the resource (in addition to being able to assign the resource to a department or to the plant)[92]. This functionality is called a Capacity Interval in Galaxy APS, and each Capacity Interval can be copied from a previous one, and be easily changed slightly to create a new Capacity Interval. You must choose if you want the Capacity Interval to be assigned to just one resource, to a department (a sort of artificial grouping of resources), or to a specific plant, or to all plants.

[91] Petroleum Refining in Nontechnical Language

[92] Departments do not have any effect on the production schedule; they are groupings of resources that often correspond to the ERP work center.

The fact that the Capacity Interval is assigned to any number of resources means that the capacities of large numbers of resources can be updated by changing just one Capacity Interval. This is a far more flexible arrangement than exists if the operating times are assigned directly to the resource, and means that just a few capacity intervals can drive all the resources in the system. This is why the approach that PlanetTogether has taken with the data is so low maintenance.

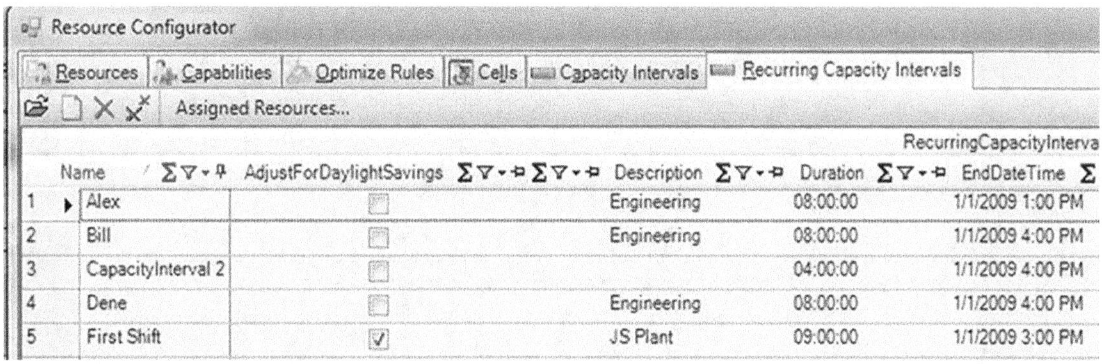

Different Capacity Intervals can be created for the normal operating time but then extra Capacity Intervals can be set. For instance, an overtime Capacity Interval could be created for one, several, or all factories from 12:00 AM to 1:00 AM, allowing another hour of production but at an extra cost. This would allow the production to continue, while reflecting the reality that the production/process order is now cutting into maintenance time and may incur more costs during the plant downtime.

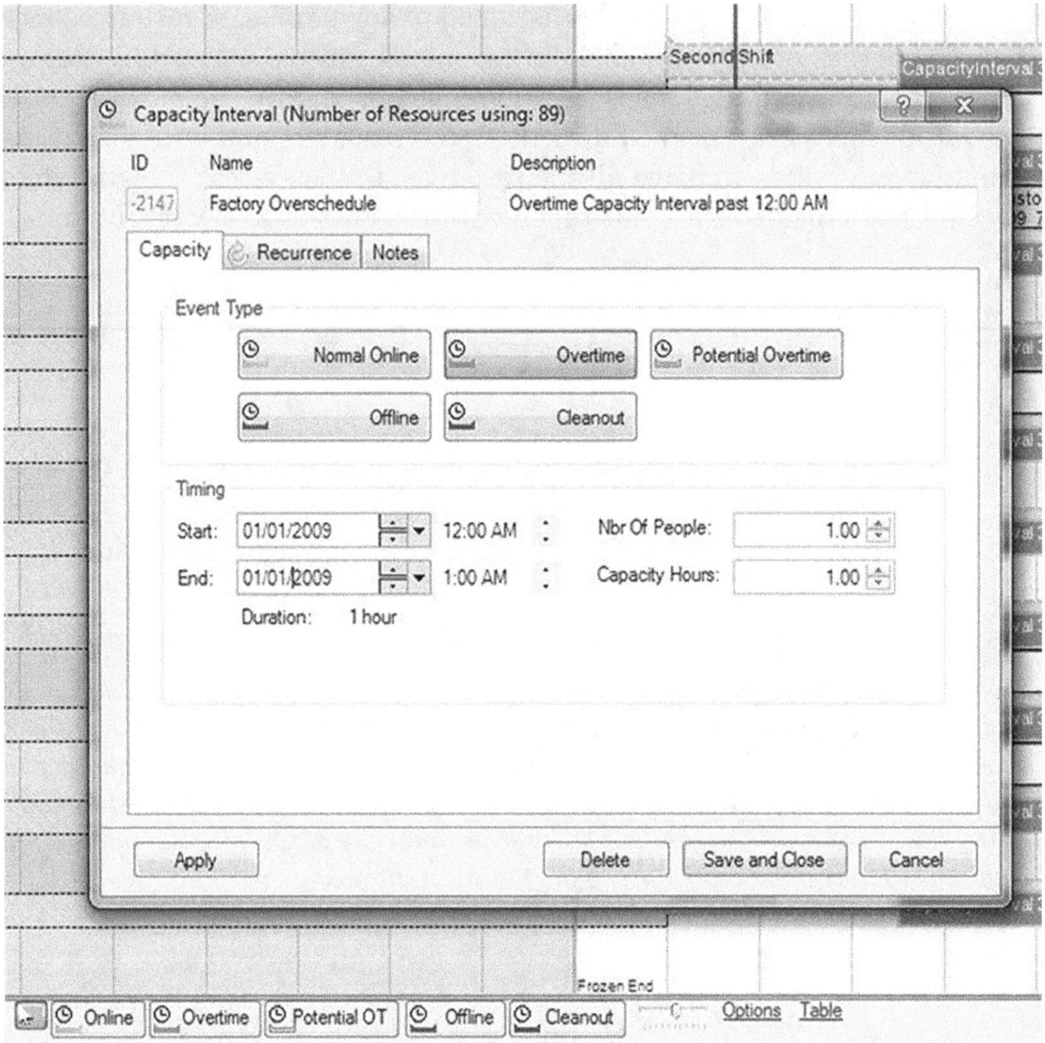

This Capacity Interval is being set as an overtime interval, and as such will have extra costs associated with it. Notice that the time is being set for one hour past the 12:00 AM plant close time, but it could be set for whatever figure is acceptable to the plant. At some point, there is a hard stop and the plant must shut down for the night. The Capacity Interval should be set out to that time. It should be noted that this Capacity Interval only has to be created once, and it can be applied to as many factories (or even all the factories) as desired. Other timing constraints can be taken into account with different Capacity Intervals, and multiple Capacity Intervals may be assigned to any number of factories, departments, or individual resources.

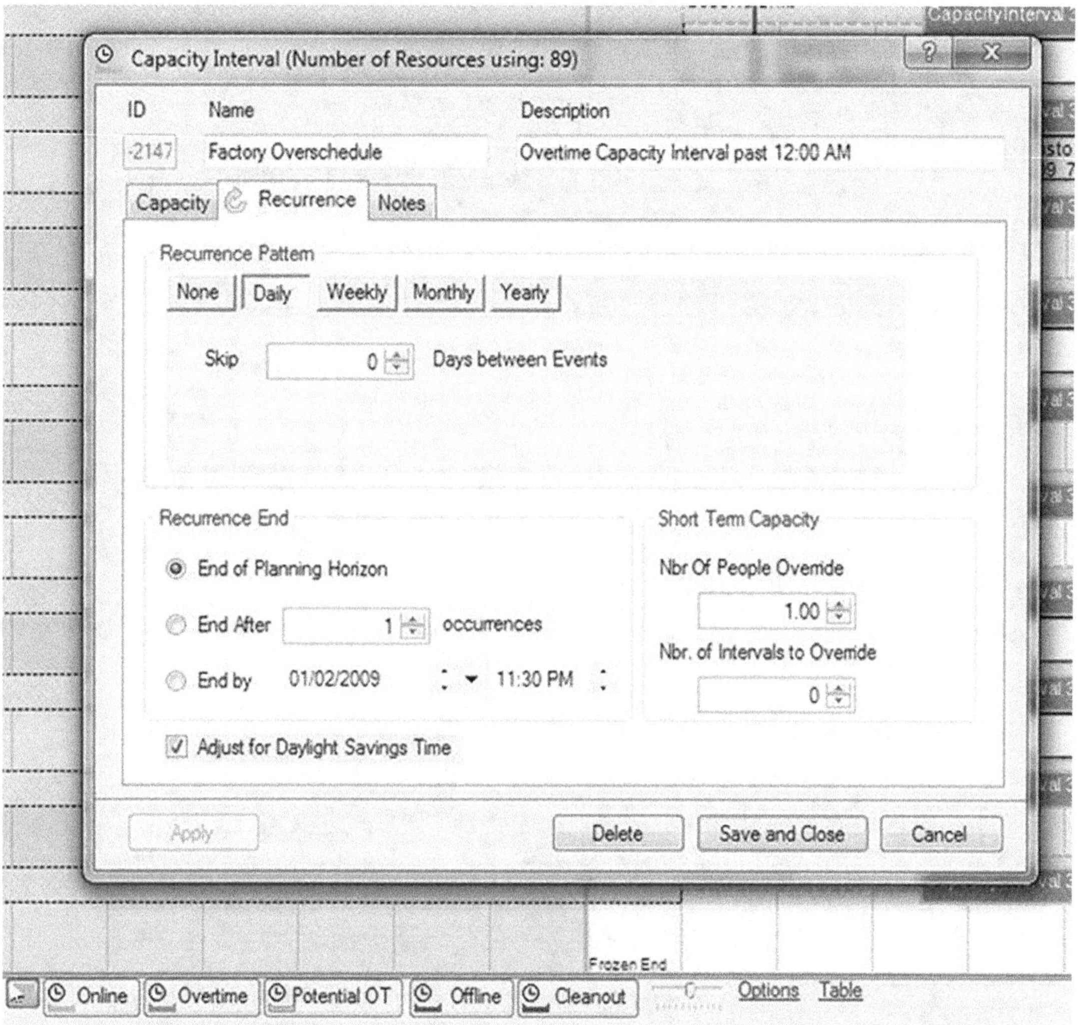

Finally, the recurrence tab defines the frequency the Capacity Profile will be used. The frequency is completely controllable and even one-time Capacity Intervals can be created.

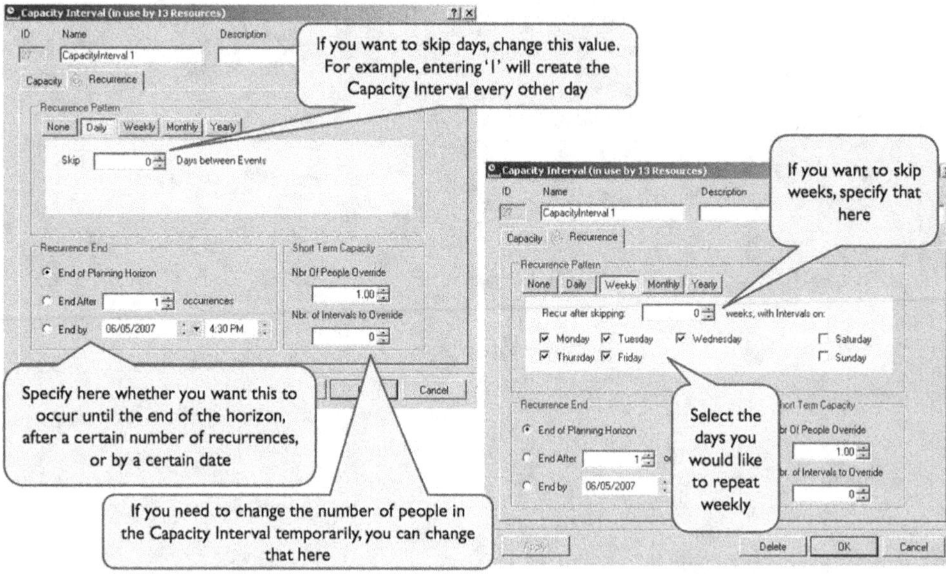

This screen shot shows the many options that are available to adjusting a Capacity Interval.

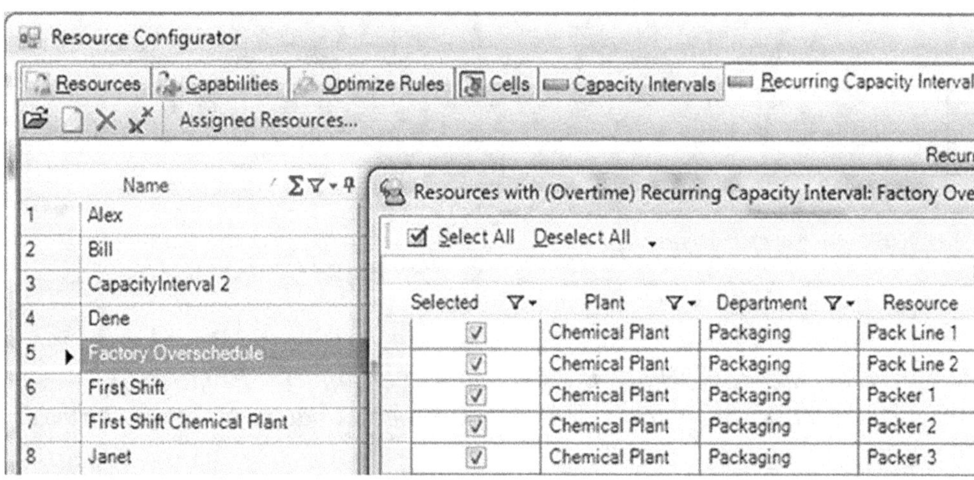

After the Capacity Interval is created and saved, it can be assigned to the appropriate factories in the Resource Configurator by selecting "Assigned Resources" at the top.

All the factories/plants, departments and resources appear in this view—not just the applicable ones assigned to the resource.

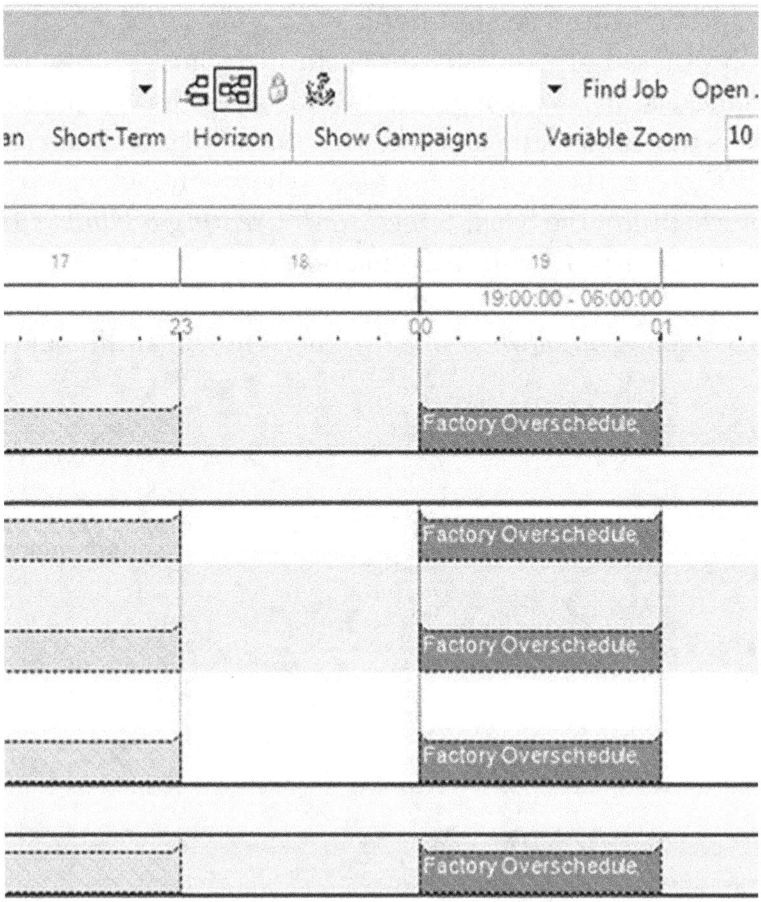

The plant overschedule immediately appears within the Gantt chart view as a red block, as opposed to the green "Normal Time" block that precedes it. Currently, there is a gap between the Normal Time Capacity Interval and the Factory Over-schedule Capacity Interval, which ends at 11:00 PM, not at 12:00 AM.

In real life, the two Capacity Intervals would be right next to each other, allowing the production run to exceed the Normal Time—but at an extra cost. PlanetTogether also has the ability to set Capacity Intervals Profiles for cleanout, which is specifically designed for the process/food processing industry.

Product-specific Resource Assignment

Products are processed or are valid for certain resources. How the application controls the relationship between products and resources is often referred to as a ***routing***. This is the "route" the materials/products take through the plant (actually in PlanetTogether, the routing can also go between factories; this is a topic I cover in greater detail in the book, *SuperPlant: Creating a Nimble Manufacturing Enterprise with Adaptive Planning Software*).

The following screen shot shows a resource assignment in PlanetTogether.

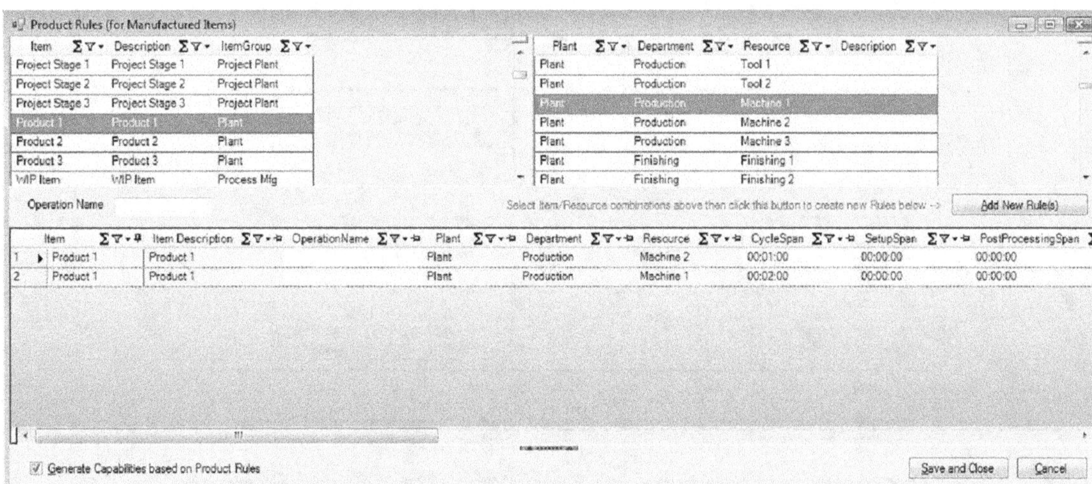

Notice from this screen shot that it's easy to assign the product to resources.

PlanetTogether describes the importance of this functionality in the following way:

> *"In any manufacturing process there are limitations on which machines are capable of producing which products: machine A may be able to produce products x and y while machine B can produce products y and z. Assuming machines A and B are in the same work center, this can be handled by simply specifying the required work center in the product formula/routing for each production step. In the Food Process industry this method of work assignment can*

be much more complex as more flexible logic is needed." — Using PlanetTogether for Effective Scheduling in the Process Industry

Physical Properties Limitations

A physical property constraint limits the products that can be scheduled on the machines. For instance, there may be multiple grinders in a plant, but some of the meat in the plant can only be ground on specific grinders. In the example below, some meat grinders can process some products but not other products.

Meat to Grinder Relationship

Grinder	*Pork*	*Beef*
A	X	
B	X	X
C		X

In this example, grinder B can grind either pork or beef. PlanetTogether explains physical property capabilities the following way.

> *"In addition to treating the work center as a required Capability, additional physical property Capabilities can be defined for the product. PlanetTogether will only schedule a product on a machine if it has all of the Capabilities required. Thus, in our example, a product may be a beef product and so only machines B and C would be considered eligible since they are grinders that can grind beef.* — Using PlanetTogether for Effective Scheduling in the Process Industry

This is managed through the assignment between the product and the resource.

Product Rules in PlanetTogether

An important part of process industry manufacturing is the ability to alter the production rates and yields of different operations, which means it is necessary to model the variability in process manufacturing. Let's go back to the example of the milk processing company, if you will recall the characteristics of the milk included the amount that it had rained, the time of year, along with a host of other factors that impact the percentage of the output product (milk, whey, curd, pharmaceutical fillers) that could be produced for a given quantity of input product. Thus the yield of different input products changed the finished goods output.

The Product Rules table can be used to define varying production ***rates, yields and preferences*** for specific items produced as Products by Resources. These values override the standard values set in the Job Operations. Rules are created for different product, plant, department and resource combinations.

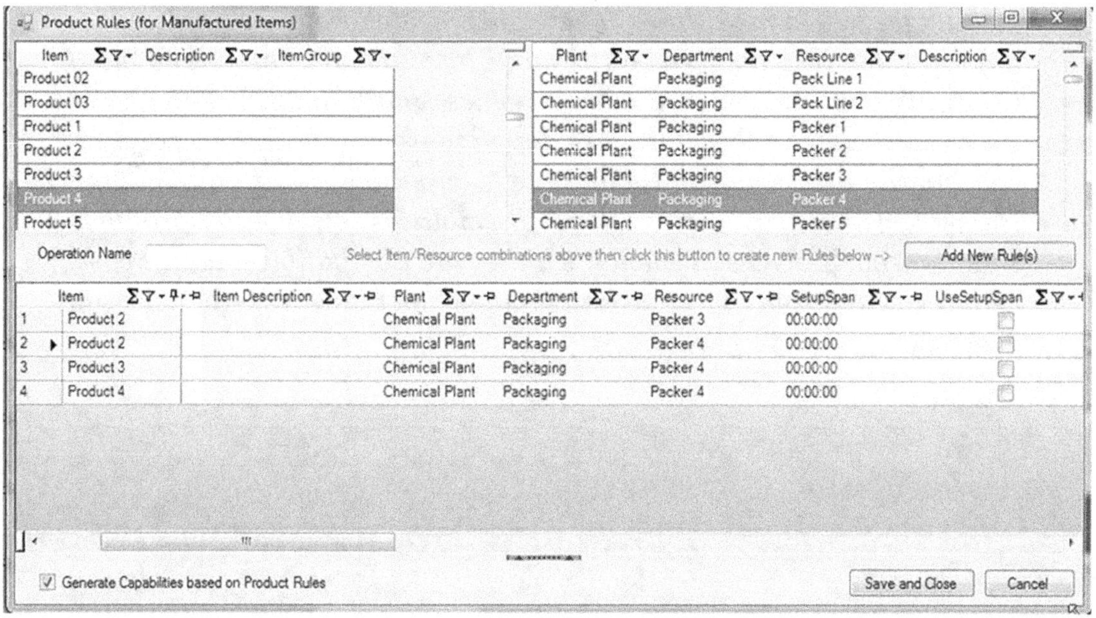

In the example above, a Product Rule has been created for the combination of the Chemical Plant, in the Packaging Department, for the Resource 3. This is a valid relationship that will adjust the production rates, yields, and/or preferences for this combination.

Compatible Scheduling Groups

Some resources require that the material be ***compatible*** with the materials being run on other resources. For example, if the same material input pipe feeds two machines, then at any point in time the machines can only run products that use the same material. Having the ability to model this production floor reality is very useful for process industry manufacturing where compatibility constraints are quite common. One of the better examples of this is food allergy constraint. PlanetTogether describes Compatibility Constraints in the following way:

> *"In multi-stage food production there are situations where an intermediate is produced by a machine which directly feeds multiple other machines. The flow of material is direct (food transport may even be automatic between the equipment) and so the second stage machines are limited to producing output from the same intermediate at all points in time. For example, cookie dough might be blended in a blender, which*

directly feeds two mold lines. Therefore when chocolate chip dough is finished blending and fed to the mold lines it's not possible to mold chocolate chip cookies and sugar cookies simultaneously—the mold lines only have the chocolate chip dough to work with. PlanetTogether can handle this constraint using the Compatible scheduling feature. PlanetTogether would only allow the molders to schedule products simultaneously across the molders if they are Compatible—in this case, if they use the same intermediate dough." — PlanetTogether Lesson Manual

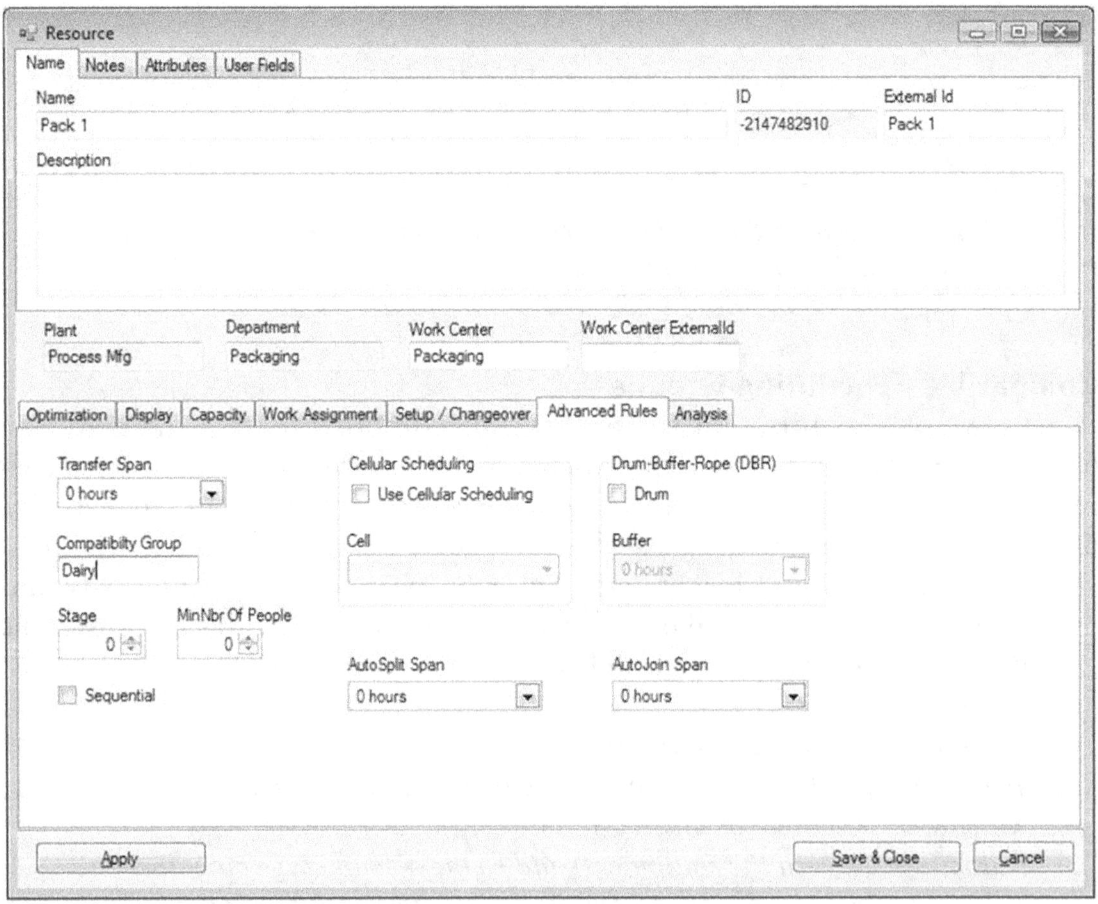

The Compatibility Group is assigned directly to the resource. A Compatibility Group is created first, and then it is assigned flexibly per resource.

All resources within the assignment must be compatible with one another. Compatibility Groups can be assigned in a way where resources may be part of one Compatibility Group, but not another. See the following example:

- *Compatibility Group 1 = Resource 1, Resource 34, Resource 90*
- *Compatibility Group 2 = Resource 12, Resource 34*
- *Compatibility Group 3 = Resource 19, Resource 22, Resource 85, Resource 110*

If these were the only Compatibility Groups, then Resource 34 would be part of two different Compatibility Groups. However, the two Compatibility Groups would not interfere with one another. Resources that are part of the same Compatibility Group can only run operations ***concurrently*** if they have the same compatibility code that is specified in the job operation.

Movements Between Operations

Many production planning and scheduling applications focus primarily on the operations that perform the processing of materials and consume most of the overall production time. However, this is not sufficient to fully model the manufacturing process because movements between operations must also be planned and accounted for. These next two modeling functionalities deal with this issue.

Connected Resources/Resource Connectors

Most scheduling applications only model the resources. Any activity that takes place between operations is also an operation, although it is not generally thought of as such. The reason for this is that the material must find its way between the two operations. If we take the example of a manufacturing process with three operations for processing 600 pounds of material, we can see why this is true.

Operations and Connections

Operation #	*Description*	*Time (In Hours)*
1	Material Prep	4.00
	Movement from Operation 1 to 2	0.75
2	Mixing	2.00
	Movement from Operation 2 to 3	0.35
3	Packaging	1.00
Total Time		8.10
Connection Time as Percentage of Total Time		14%

In this example, if we only model the "operations" we miss fourteen percent of the time that it actually takes to produce the finished product. Moving material or other material prep does not normally fit within the definition of an operation. Typically an operation is an activity that takes place on a machine/resource or is processed by people/resources. However, the approach of considering connections between operations as also being operations allows a more accurate modeling and therefore planning of the plant floor.

PlanetTogether has a separate modeling setup for this, which it calls Resource Connectors. Resource Connectors allow the modeling to be even more precise by segmenting the main operations from the intermediate steps necessary to move material between operations.

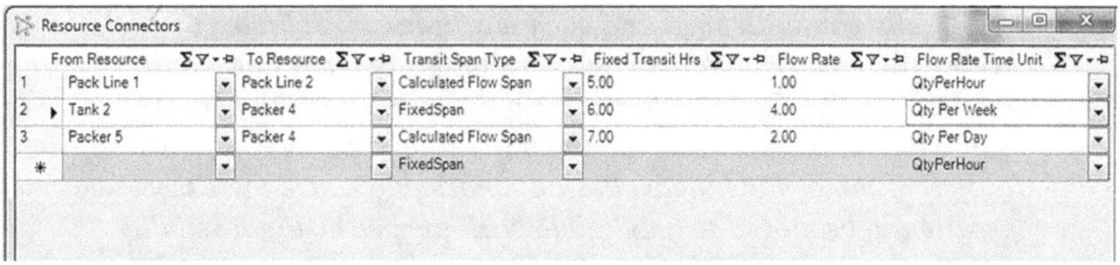

The Resource Connector configuration is shown above. It simply requires the "from" and "to" resource, as well as times and transit hours. The optimizer then takes all of these into account during the planning run. If we take the example above of Tank 2 to Packer 4, we can see that the type of Transit Span is a FixedSpan of six hours, at a rate of four per week.

Material Handling Equipment Constraints

What has just been discussed is a very specific material handing scenario that requires specific constraints.

"The movement of intermediate food product between operations is often done with the use of direct feed material handling equipment with no manual intervention. For example, molded cheese might fall directly from a molder into a transport canal where it cools while being transported to the packing workstation. This creates a scheduling constraint because the selection of machines for molding

and packaging are interdependent. The cheese can only be packaged on a pack line that is set up to service the particular transport canal. There may be other pack lines available but ***if they are not connected to the mold machine then they are not eligible to do the packing****. Galaxy APS can handle this type of constraint in a few ways.*

1. *First, 'Cells' can be defined and each machine can be linked to a cell. In our cheese example, we might create 'cell 1' and link mold line 1 and pack lines 2 and 3 to cell 1. Then Galaxy would only schedule the pack operation on a pack line in the same cell as the mold line. Moving a production run to a different mold machine would automatically cause the associated pack operation to be moved to an eligible pack line, ensuring feasibility.*
2. *If more sophisticated logic is required, there is also the option of defining a custom constraint. For example, if the material handling equipment can be moved between machines, then limiting a pack line to a particular cell may not be flexible enough. A custom material handling constraint rule can be easily created to solve the problem."* — Using PlanetTogether for Effective Scheduling in the Process Industry

Shared Input Constraints

In some process industry manufacturing, an intermediate product feeds more than one machine that follows in the sequence. In some cases, this product feed is automated and continuous. When this is the case, the following resources must process the same intermediate product.

Shared Input Product

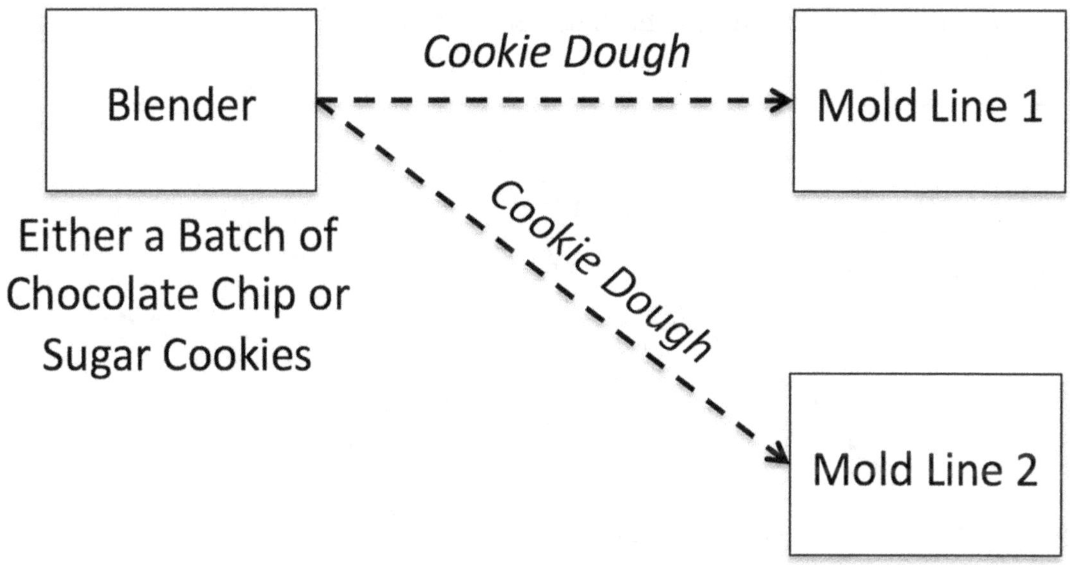

In multi-stage food production there are situations where an intermediate is produced by a machine that directly feeds multiple other machines. The flow of material is direct (food transport may even be automatic between the equipment) and so the second stage machines are limited to producing output from the same intermediate at all points in time. For example, cookie dough might be blended in a blender that directly feeds two mold lines. Therefore when chocolate chip dough is finished blending and fed to the mold lines it's not possible to mold chocolate chip cookies and sugar cookies simultaneously—the mold lines only have the chocolate chip dough to work with. Galaxy APS can handle this constraint using the Compatible scheduling feature. Galaxy APS would only allow the molders to schedule products simultaneously across the molders if they are Compatible—in this case, if they use the same intermediate dough. — Using Galaxy APS for Effective Scheduling in the Food Processing Industry

AspenTech and a Crushing Operation

Up to this point we have primarily discussed AspenTech in terms of its ability to model oil and gas processes, however, it can apply the same customized resource standards to quite a few resources types. For instance, AspenTech can estimate the efficiency of a common process operation that is used across process industry manufacturing types, such as the crushing example below:

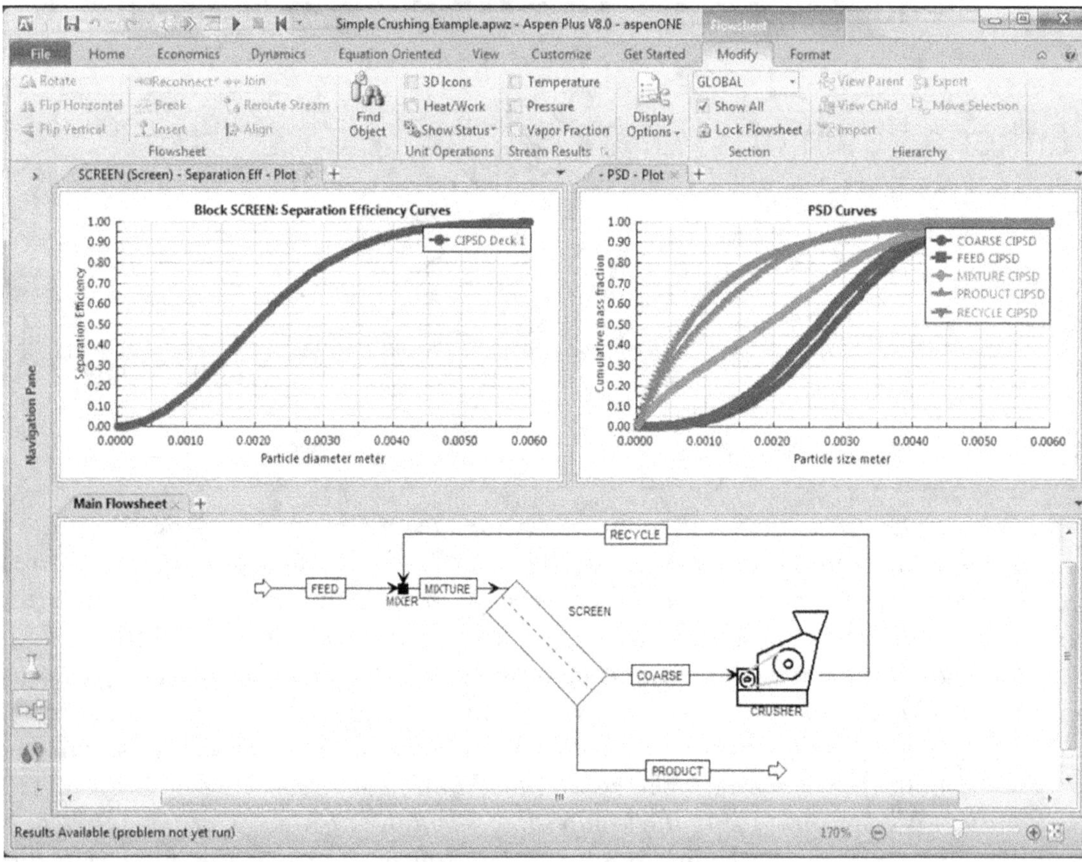

Notice that estimations for the particle diameter and the particle size are shown side by side. This type of functionality is simply not available in more generic production planning and scheduling applications.

These statistics are determined by how the crusher is configured.

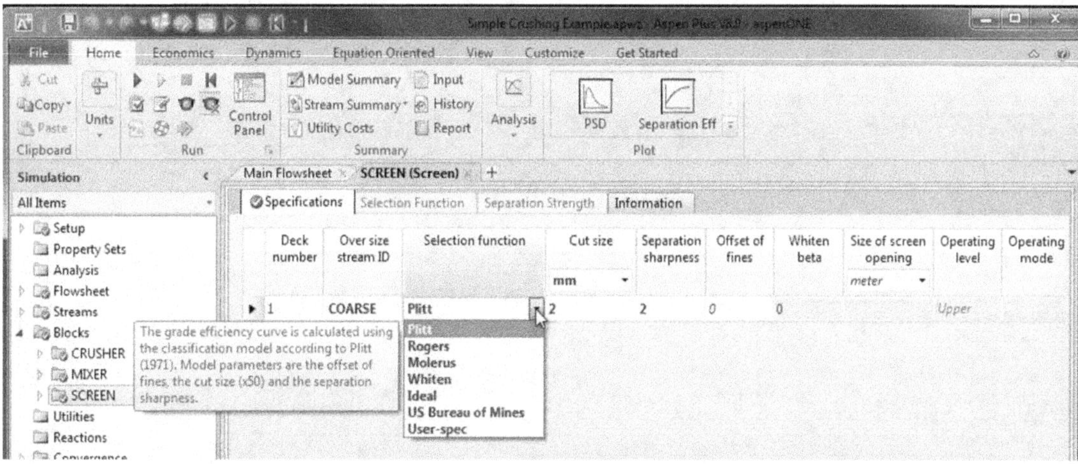

Notice that the crusher is defined by its function. There are several crushing functions to choose from, and each crushing function has a different set of mathematics, which will control the estimation. In this case, the Plitt classification model is employed. The fact that there are so many functions to choose from really shows how this application can be customized to the task.

It is also defined in terms of the cut size (with a unit of measure of millimeter), the sharpness of separation, etc. These options are available because we have selected to configure a crusher. However, AspenTech has separate configurations for different resource types. When compared to any number of production planning and scheduling applications this configuration is easy to understand. This high customizability for resources combined with ease of configuration provides a highly flexible modeling environment.

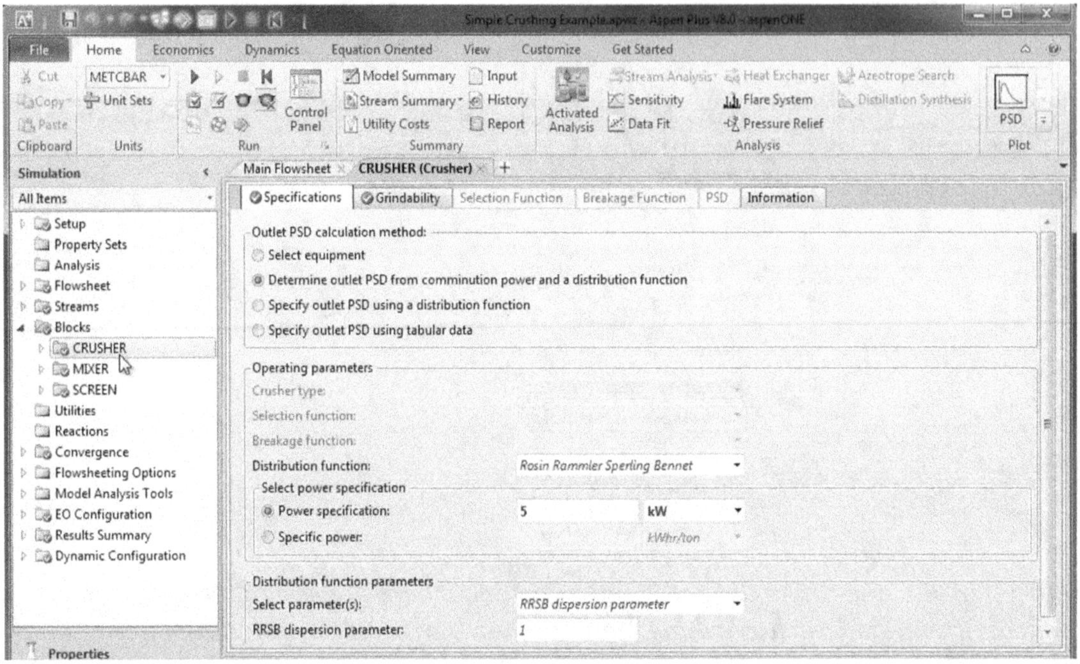

More specific details are configured in the Specifications tab where the power of the crusher is defined, as well as the crusher's distribution function. The option at the top of the screen tells the application to use both comminution power and a distribution function (which is then defined below). Again, this crusher is being defined in a very specific and detailed way that acknowledges the operating rules of the crusher.

Crusher's can be just one set of resources in a larger manufacturing process within a plant, or they can be the process of the entire plant.

Conclusion

Constraining manufacturing resources that do not change frequently is one thing, but a major complexity of process manufacturing is the variability of its constraints. Variability is frustrating to model because specific output quantities cannot be depended upon. Part of the job of process manufacturing is to control this variability (this is referred to as "process control") and this is performed outside of the planning system and there are steps designed to reduce the actual variability in the factory. The computerization of process control panels not only provides better and more flexible monitoring of the plant, but also for reviewing the historical data, which due to computerization, can now be archived and mined for insights. However, even with process control, variability still exists, and process industry manufacturing tends to have the highest variability of the different manufacturing environments. There are several different sources of the variability. One source is the input material, called the feedstock. A second source of variability is in the yield of the output. While discrete and repetitive manufacturing has scrap, the output variability is nothing like the yield variability of process industry manufacturing. Quite an important measurement of any production planning and scheduling system is how well resource variability can be modeled. Most people who work in production planning are familiar with how capacities are set on a resource in whatever application they use. However, it can also be important to set minimum limits, so that a resource is not used unless it can run a minimum amount of product through for a planned order. The fact that planning departments are run with the absolutely minimum staffing level, means that companies must do a better job of finding low maintenance applications—and this can only be determined by testing the application on important maintenance criteria, one being how well the application can account for all forms of variability. There are, in fact, other types of variability that are not discussed in this book but relate to the application being able to be switched between sources of supply among the internal production locations, and among external production locations such as subcontractors and contract manufacturers. This is the superplant concept and the functionality to account for this variability is covered in the SCM Focus Press book, *SuperPlant: Creating a Nimble Manufacturing Enterprise with Adaptive Planning Software.*

AspenTech is a leader in several process industry areas, and most prominently in oil and gas. This chapter explained how their highly specialized software can be used to model oil and gas processes, to simulate in fine detail a planning run, and can allow the planner to made adjustments to feedstock and to production settings in order to bring in the finished good on the correct specification and in the correct quantities.

PlanetTogether provides functionality in the Capacity Interval that can be flexibly assigned to resources. The fact that the Capacity Interval is assigned to any number of resources means that the capacities of large numbers of resources can be updated by changing just one Capacity Interval. The assignment between resources and Capacity Intervals can be quickly changed either within the resource view or within the Capacity Interval view.

An important part of process industry manufacturing is the ability to alter the production rates and yields of different operations, which means it is necessary to model the variability in process manufacturing. In Galaxy APS the Product Rules table can be used to define varying production ***rates, yields and preferences*** for specific items produced as Products by Resources.

Most scheduling applications only model the resources. Any activity that takes place between operations is also an operation, although it is not generally thought of as such. The reason for this is that the material must find its way between the two operations. If we take the example of a manufacturing process with three operations for processing 600 pounds of material, we can see why this is true. PlanetTogether has a separate modeling setup for this, which it calls Resource Connectors. Resource Connectors allow the modeling to be even more precise by segmenting the main operations from the intermediate steps necessary to move material between operations.

CHAPTER 10

Manufacturing Execution Scheduling

Manufacturing execution scheduling (MES) software covers the phase after planning and works in conjunction with the process control system(s). MES systems take up the functionality space between the ERP system and the process control systems (which are discussed in Chapter 9: "Constraint Planning and Process Control." Many people are not quite sure why an MES is necessary and some casually assume that ERP systems provide what is necessary as ERP is an "execution system" after all. This question is well answered in the following quotation:

> *"True, the ERP system does create a production schedule, but it isn't yet tailored to the actual capacity of the production lines, nor has it taken efficiency into account. That's why in nearly every factory; you'll find a scheduler who plans in Excel. These are people who carry around a great deal of knowledge in their heads. They really mustn't ever get sick or go to vacation, and they feel tremendously valued. The production department is completely dependent on the person. In the meantime, the plant manager's asking herself*

> *who's really the boss."* — MES Guide for Executives: Why and How to Select, Implement, and Maintain a Manufacturing Execution System

MES systems tend to be applied most frequently in the process industry manufacturing sub-categories where tracking is very important for regulatory compliance. A perfect example of this is pharmaceutical manufacturing. In fact the motivation to implement MES systems for pharmaceutical companies may be as much related to reducing paperwork and increasing traceability as they are for the ostensible first purpose of MES systems, which is to improve the efficiency of manufacturing execution.[93] A second, and more primary functionality provided by MES systems is the ability of anyone with the appropriate privileges to access the data accumulated by the MES system, and therefore to understand what is actually happening in the plant. This ability extends to both remote as well as on-site personnel. Furthermore, because so much manual reentry is eliminated with an MES system, the information is more accurate not only because of keying issues but because MES systems perform and essentially automate the calculations that are previously performed by plant workers—increasing the accuracy of the calculations that need to be performed. In fact, much of the tedious and error prone work is removed from the manufacturing process when a successful MES application is implemented.

The MES and Inventory Accuracy and Visibility

MES software is also expected to improve inventory visibility. This is particularly important for process industry manufacturing because of the limitations that most ERP systems have in properly reflecting inventory of finished goods that do not yet have an item number assigned (as they have not been graded). Because of how ERP systems view and record inventory, it makes much more sense to use the MES as the **real time** inventory information system, and then to update the ERP system when "things have been worked out" on the manufacturing floor. Interestingly, in a study by AMR (although it was from 2005), only 59 percent

[93] This is certainly understandable when it has been estimated by the MES Association (MESA), that between 50 percent to 80 percent of paper batch records contain errors of one type or another.

of respondents of 114 pharmaceutical manufacturers has automated updates of inventory data from the manufacturing system to the ERP system.[94]

Because MES systems deal at such a detailed level of information, companies that have different plant types will tend to have more expensive MES implementations. In fact, in this situation, an MES system, which is a good choice for some of the plants, may not be the best MES for other plants. As is covered at research at our companion site Software Decisions, the overall cost reduction, which will come from using an existing application, averages out to be only roughly one quarter of the total cost of ownership for the life of the application (unless the application is a Tier 1 ERP, in which case the numbers are higher).[95] Because of this, it is a simple matter to exceed this cost savings with an increased likelihood of the following:

1. A Longer Implementation
2. More Customization Expense
3. A Higher Risk Implementation
4. Lower Functionality/Worse Fit of Functionality
5. Lower Usability
6. Lower Maintainability

These issues generalize across all of the enterprise software categories. Yet companies continue to receive the opposite advice from their consulting companies—that is to choose one application and customize it—or make it fit, and this advice is given in many cases because the consulting companies have resources that are trained in a particular application.

MES systems contain the following types of data and activities:

1. Resource Status
2. Operations Scheduling

[94] Realizing ERP's Untapped Potential
[95] Software Decisions is at http://www.softwaredecisions.org

3. Production Dispatching
4. Document Control
5. Data Collection
6. Labor Management
7. Work in Process Status
8. Quality Management
9. Performance Analysis
10. Process Management
11. Product Tracking and Genealogy
12. Maintenance Management[96]

Two well-known vendors in the MES space are Wonderware and AspenTech. AspenTech covers both planning and MES, as well as a number of other software products.

[96] Manufacturing Execution System

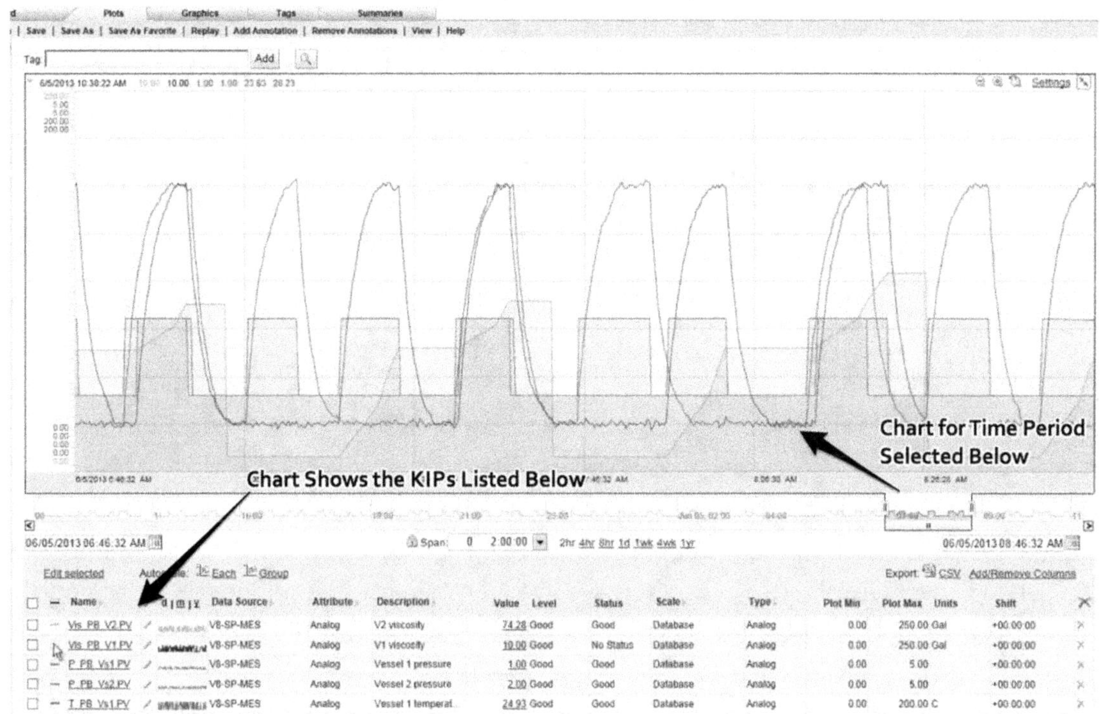

AspenTech's MES product shows manufacturing KPIs in great detail, allowing the process control user to zero in on variability and to make changes quickly. AspenTech's MES product can be used on a computer, tablet or mobile device—bringing up interesting applications for remote monitoring.

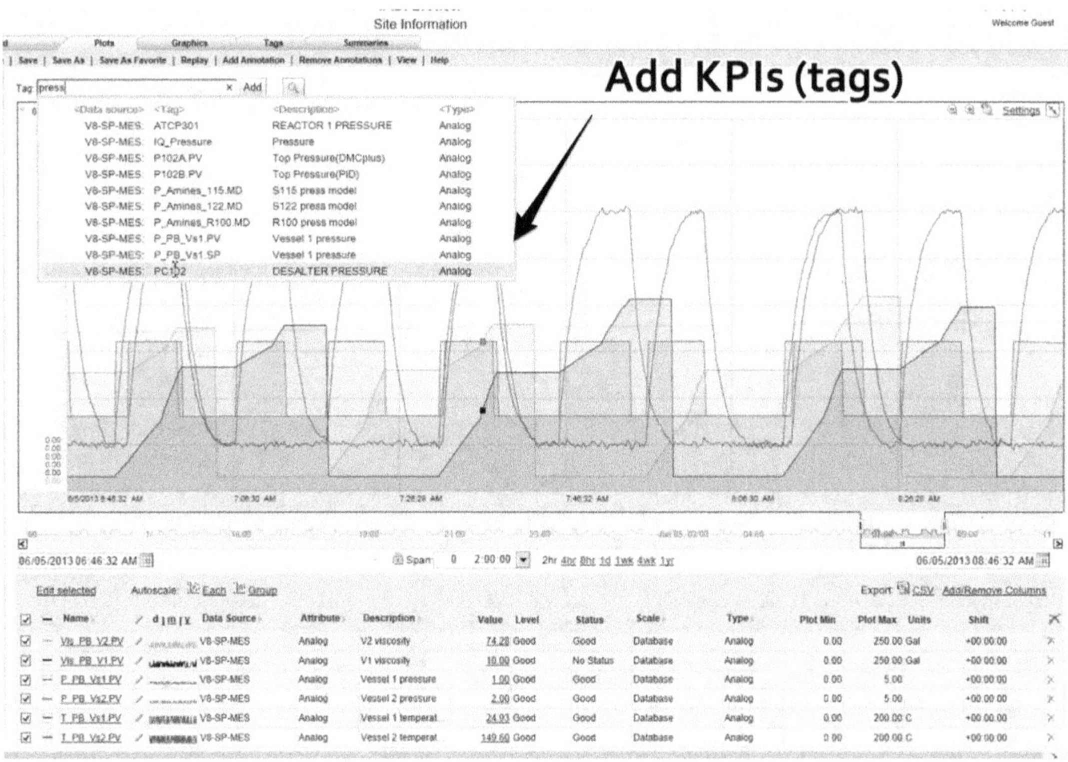

The KPIs that are tracked, tags in AspenTech, can be added by simply searching for the tag database in the upper left corner and then adding them.

MES and The Data Historian

Data historian software is most commonly found in large-scale continuous manufacturing environments. Wonderware is one well-known software vendor that offers data historian software. Data historian software is a non-relational database technology that holds vast amounts of time series data and uses SQL. Data historian software is a database that is optimized for input-output and for real time data gathering. While it sounds similar, this approach as well as the data model is quite different than simply a business intelligence solution because data historian software is designed to be a real time information system that means being constantly updated. Many software vendors that offer MES software (as does Wonderware) also offer data historian software. The data historian software offered by the same vendor that provides MES will naturally often interoperate with the MES.

Déjà vu for the Process Manufacturing Industry

Of all the manufacturing software described and explained in this book, MES applications are the least widely implemented and are also the least mature and have the fewest competent offerings. A primary reason for this is that most companies use home grown shop floor applications, and the detailed level of information that is required in the plant, is so specific that it has been a frequent area of custom development rather than off the shelf application purchases. Furthermore, one of the confusing things about MES software is that the vendors propose so many potential areas of improvement that can hypothetically be provided to companies that implement an MES system. MES applications also have considerable overlap with other manufacturing applications. For instance, in process industry manufacturing, the MES has some overlap with process control software, which is far better entrenched in the industry. And that is just one area where there is overlap—other areas are with the ERP system as well as with planning systems. For instance some MES software vendors describe feedstock optimization as a role for the MES, but that certainly does not seem like "execution," and is better performed by planning systems. Secondly, many of the areas of functionality promised by MES vendors turn out to be the same areas of functionality that were promised, but **never actually materialized** from ERP vendors. For instance if one reads through the list of functionality of an MES vendor.

- Inventory Accuracy
- Costing and Yield Prediction
- Managing Variable Product Characteristics: *"The effect of variable product characteristics and inventory attributes, such as package weights, expiration dates, and multiple units of measure, on inventory management, order management, production scheduling, manufacturing, QC management and product costing."*[97]
- Ingredient Tracking
- Quality Tracking

One can be forgiven from observing that all of these were previously promised with ERP systems! For many years ERP vendors held out the promise of improving manufacturing operations—and the result is a steady stream of criticism

[97] Impact of Formula-Based ERP Applications on Pharmaceutical Manufacturers

that ERP systems do a poor job both in managing the factory and in connecting to shop floor systems.

This is quite interesting, and made all the more interesting that the ERP promises in this regard are not highlighted as not having been met before analyzing MES systems. This would seem to be a natural starting place for analyzing both the existing ERP system and MES systems, however as is covered in detail in the SCM Focus Press book *The Real Story Behind ERP: Separating Fiction from Reality*, there is a constant amnesia when it comes to holding ERP vendors accountable for what they promised so confidently during the sales process.

What this all means is that proper implementation of an MES really requires a solution architect with a strong knowledge of all the process industry manufacturing capabilities, knowledgeable of multiple software vendors in each of the process industry manufacturing software categories, but one which is not beholden to any **one application category**. This type of knowledge base, in addition to restricted bias is very difficult for process industry companies to find inside their companies. It's also very hard to find this combination of skill sets in consulting companies, as they normally have preselected vendors they work with and are intent on directing their clients into before they ever review the client's requirements—and these solutions are those that make the consulting company the highest margin—not the best solution for the company.

CHAPTER 11

ERP for Process Industry Manufacturing

Unfortunately for process industry manufacturing companies, the vast majority of ERP systems have been designed for discrete manufacturing, and were then adapted to meet process industry requirements—but as with most software—applications tend to work most smoothly for the first requirements the application was designed for, as is explained in the following quotation.

> *"The Food and Beverage Industry has always confronted many challenges when it comes to business management software—from Manufacturers and Producers to Wholesale Distributors, Importers and Retailers—across the entire Supply Chain, businesses in the Food and Beverage Industry face complex requirements, which until now, have been flat out ignored by mass market software producers. As a result, many organizations in the Food and Beverage Industry have been forced to either build highly customized ERP software systems—either in whole or part—or suffer with inefficient processing to*

> *accommodate the lack of functionality available in their current software suite."* — Food and Beverage Distribution & Manufacturing Software

Therefore, it is important to level set with how process requirements differ, and to that end the following is a list of some of the major areas of difference.

1. Units of measure (UOMs)
2. Partial UOMs
3. Parallel UOMs
4. Catchweight
5. Batch Management/Lot Control
6. Recipe and Formula Management
7. Assigning Completed Production Prior to QA
8. Process Industry Manufacturing Costing
9. *Adjustment for Feedstock (necessary only if the company does not have a planning system)
10. *Master Scheduling Output (necessary only if the company does not have a planning system)

And now we will discuss each area in more detail by describing an ERP software vendor called ProcessPro that is highly focused on process industry manufacturing.

Partial UOMs

Unlike discrete manufacturing, process industry manufacturing has partial UOMs that must be tracked. Furthermore the valuation UOM is not necessarily equal to the actual unit.[98]

[98] Process Industry ERP Requirements

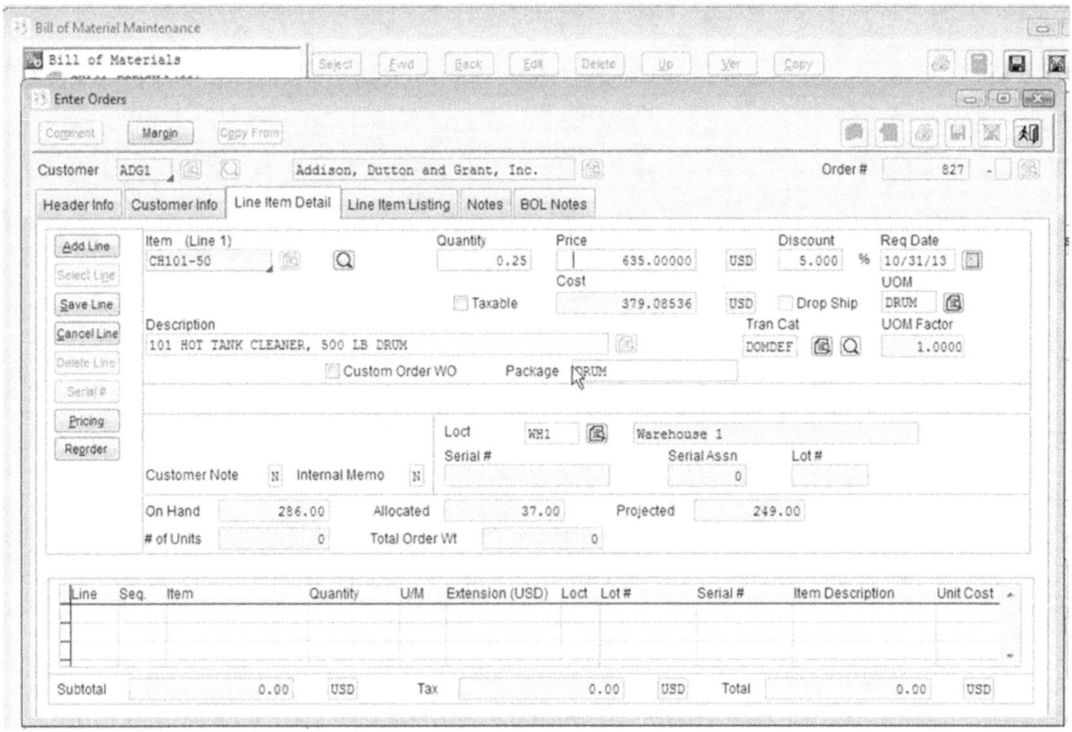

The UOM value is open by privilege. ProcessPro allows the sell factor to be a different value.

Parallel UOMs

In process industry manufacturing there are multiple UOMs, which must be accounted for.

> *"The most effective approach for handling concentrations is to support two parallel UOMs. For example, the first UOM (volume or weight) is tracked for purposes of reporting physical quantities on hand and inputting quantities to a specific process. The second UOM (concentration) is tracked for purposes of calculating physical quantities required in a specific formulation, as well as in determining purchase quantities and product costs. This type of design approach only exists in ERP systems intended for process manufacturing."*[99]

[99] Process Industry ERP Requirements

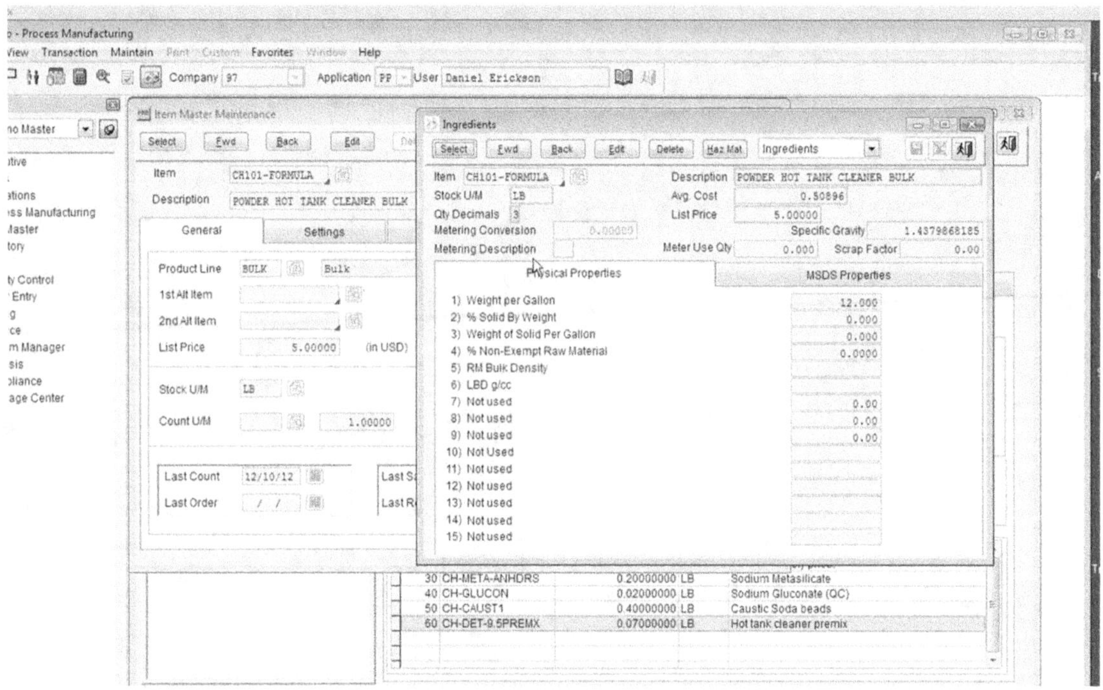

Here ProcessPro shows that it can easily hold multiple UOMs. Notice there are many UOM types that are not used, but could be used.

Catchweight

Catchweight is where the same SKU can have different weights even though the SKU is priced per pound. This is not true of discrete manufacturing so the concept of a catchweight does not apply.

> *"For example, a case of whole chickens may contain multiple chickens with an average case weight of 50 pounds. As the weight of each individual chicken may vary, the total case can weigh either over or under the standard case weight. Therefore, tracking one case out of inventory and then billing a customer for one case will not be accurate enough. To correctly value the inventory, as well as bill the customer accurately, it is necessary to also capture the total weight of the chickens within each case."*[100]

[100]Process Industry ERP Requirements

Catchweight functionality is essentially all about representing two weights through to the inventory.

> *"And by now, most food distributors who sell either boxed beef, poultry, fish, produce and cheese have discovered that catch weight functionality has been largely ignored by the mass marketed software community."*[101]

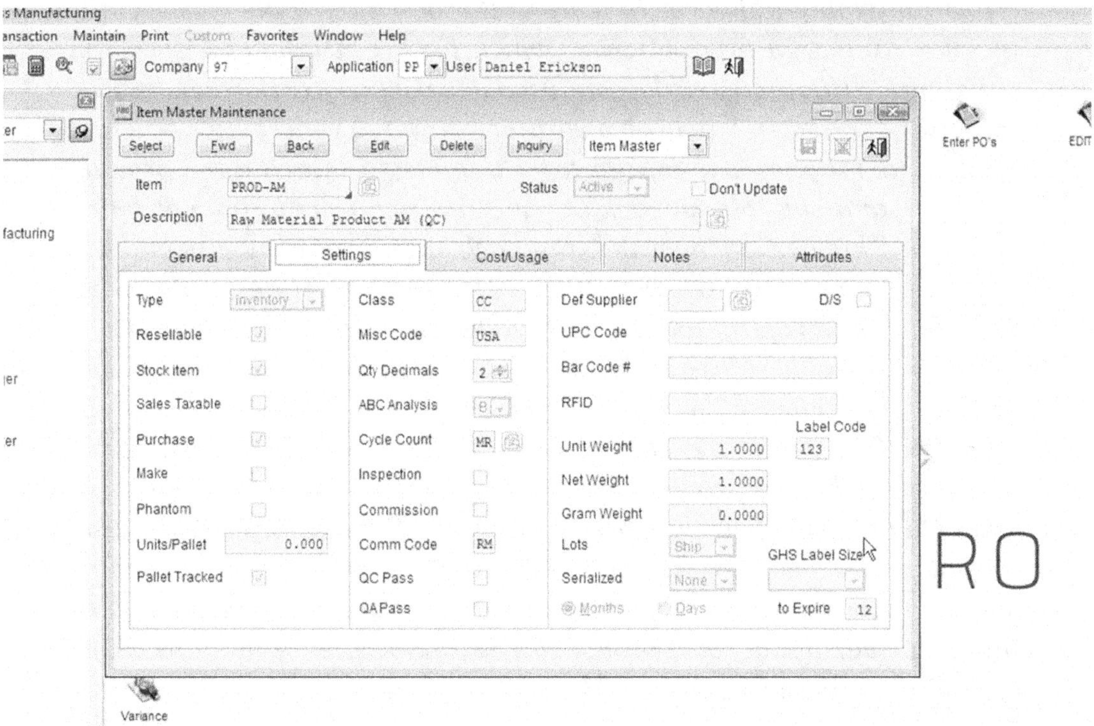

Integrate to a weight scale to capture the weight of the product. Gram weight is utilized for nutritional label. Gram is used as the common denominator for Genesis R&D Nutritional Labeling integration.

Scale weight/Unit weight can be used as the catchweight. This is a very specific requirement for specific customers. On the raw material side they are common for meat products.

[101] Catch Weight Software

Batch Management/Lot Control

There are actually several reasons for the necessity of batch management/lot control. One is traceability for safety and regulations—common both in food and beverage and pharmaceutical industries. Another reason is because of quality differences in the raw materials, the intermediate products as well as the finished goods.

> *"Splitting of lots into sub-lots. Quality—the chemical and physical properties—may fluctuate over time during production. This is especially true for campaign production in which long, continuous production runs cover differing climatic conditions as well as fluctuations in raw material quality. In batching operations running consecutive batches, operations may wish to assign a lot number to indicate the production run, and sub-lots to track the quality from time to time or batch to batch. This assists them in tying back specific quality attributes to operating parameters tracked in the MES systems. Rules and restrictions for the storage of lots/sub-lots. Whether lots can be combined or not, there are also rules about where lots can be stored (e.g., which vats in the brewery) as well as what kinds of containers the product can be stored in (e.g., product A can be stored in 55 gallon drums, but nothing larger, whereas product B may be stored in any sized container). With this information the user can access all necessary information and answer relevant questions. How much inventory of a particular specification is in stock? What is the master formulation? How did we make this particular lot? What should it have cost us? What did it cost us? What is its selling price?"*[102]

[102]Process Industry ERP Requirements

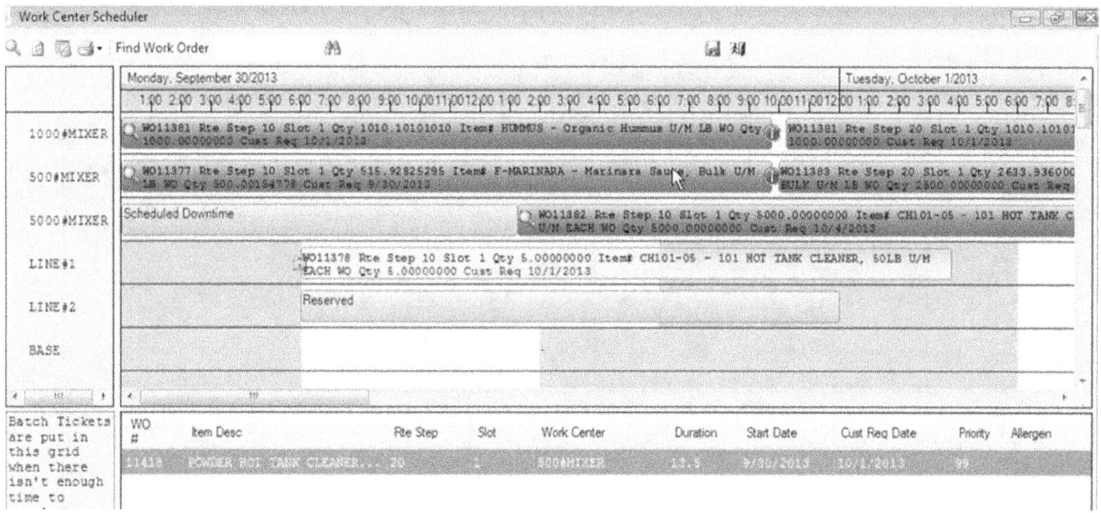

Product is produced in a batch and scheduled on the planning board. The manufacturing batch is then traceable throughout the entire inventory management and sales process. Many customers are pleasantly surprised that it can be done right in the ERP system. Some customers are using spreadsheets to deal with the batch management functionality. Capable batch management functionality in ERP systems is quite rare. All that is necessary is to denote the common batch size in the master data—and ProcessPro functionality will account for it from there on out.

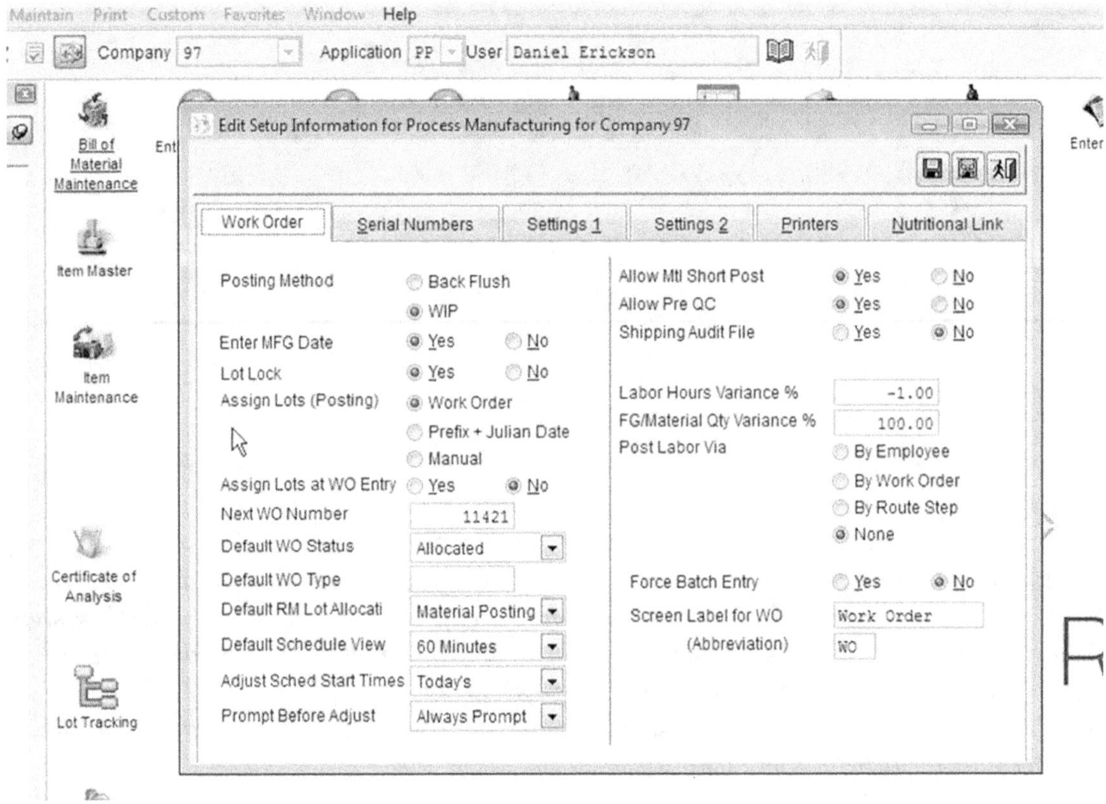

Lot Lock disabled connects one lot to one production order—which is of course a batch or group of batches. With Lot Lock enabled, the number of lots produced per production order or batch becomes completely adjustable by the production manager. With Lot Lock disabled, it assigns a lot number per posting completion event. Each company will choose how to complete.

In inventory management, batch management can be called lot size. In some customer environments a batch can be the equivalent of a lot, but in others a batch can connect to multiple lots. For one production order, say for a total order quantity of 10,000 and a production constraint (ex: vessel capacity) of 2,000 representing 5 batches to complete the order, and therefore 5 potential lots or 5 batches posted to a single lot. The length of the production run or batch will often determine how many lots are created. If a production order has a long duration (think multiple days/weeks)—then one might configure ProcessPro to have one lot per batch or day of production. Breaking up the production order into multiple lots greatly enhances traceability, and is quite useful in conditions related to recall. ProcessPro can be configured in either way.

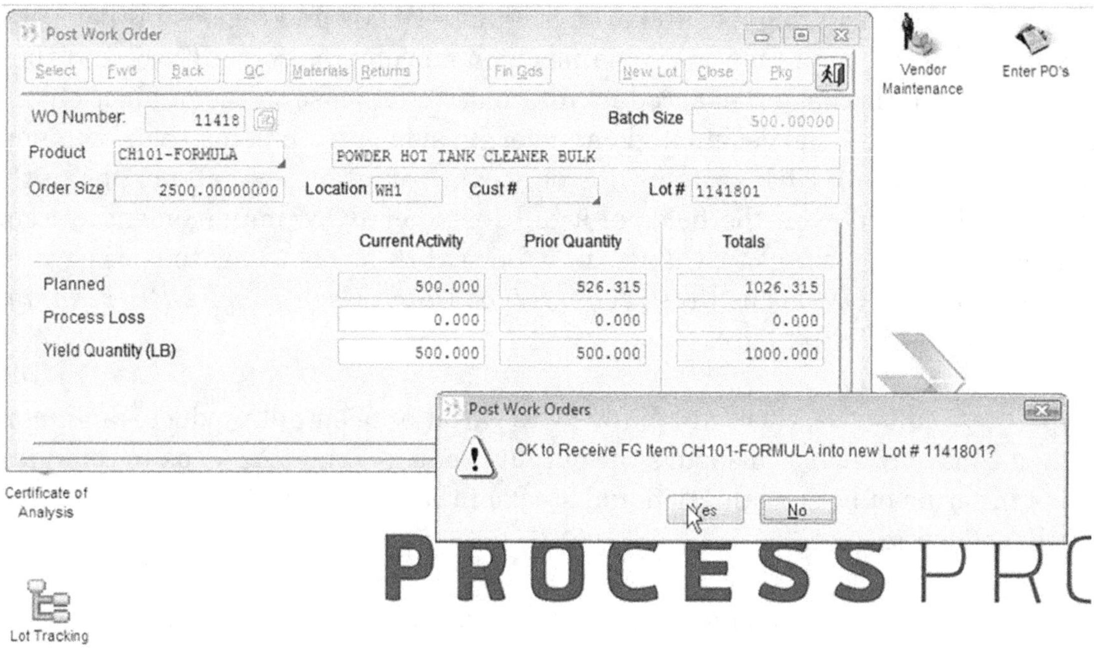

Lots in ProcessPro are determined by the production manager. A new lot can be created at any time—this is particularly useful when there is a change in the production process—and it allows a change in tracking exactly at the time when it is most useful and represents accurately the change in the production process. However, creating lots, while easy in ProcessPro, still has costs downstream (labels, new containers) and the benefit of creating a new lot must be weighed against the cost of creating a new lot.

Recipe and Formula Management

While they are often used for this, at SCM Focus we have yet to find an ERP system that performed well in recipe/formula management. There are several factors at work here. First, all that should be expected of ERP systems is to **accept recipes** and to have functionality to both accept recipes with multiple input/output products, and to allow for fractional input products. ERP systems should never be expected to originate the recipe information. This was covered in Chapter 5: "Recipe Management." This distinction between what a recipe/BOM management system does and what an ERP system does is not well understood, primarily because consulting companies have poorly explained what product data systems actually do,—and many ERP software vendors have told their clients that they can manage and even originate BOMs and recipes in their ERP system. This has led to many companies using a combination of Excel and their ERP system to manage their product information (at least on the production and supply chain side—the design and engineering side is often using their own non-integrated system). ERP systems can only ever represent a small portion of the functionality, and even the fields of a full recipe or BOM management system. This is a very important topic, but not a focus of this book. This topic is covered in detail in the SCM Focus Press book, *Bill of Materials in Excel, ERP, Planning and PLM/BMMS Software.*

In process industries, there are many more varieties of output product—and more dimensions by which to measure the output product—and this leads to complications in terms of both evaluation and stock of the output product as is explained in the following quotation.

> *"If each variant had to have a different item number (as it would using discrete ERP software) then the 'item master' would quickly become unmanageable and customers would be faced with a confusing array of 'items' to choose from when ordering. Periodic sampling of the lot determines if any of the quality parameters have changed. If this required a change in the item number on the label at the warehouse, this would mean huge volumes of non-value added work."*[103]

Therefore, recipe functionality, which is simply a slight adjustment to BOM functionality in an ERP system, will lead to complications and probably customization requirement when implemented in process industry companies.

[103]Process Industry ERP Requirements

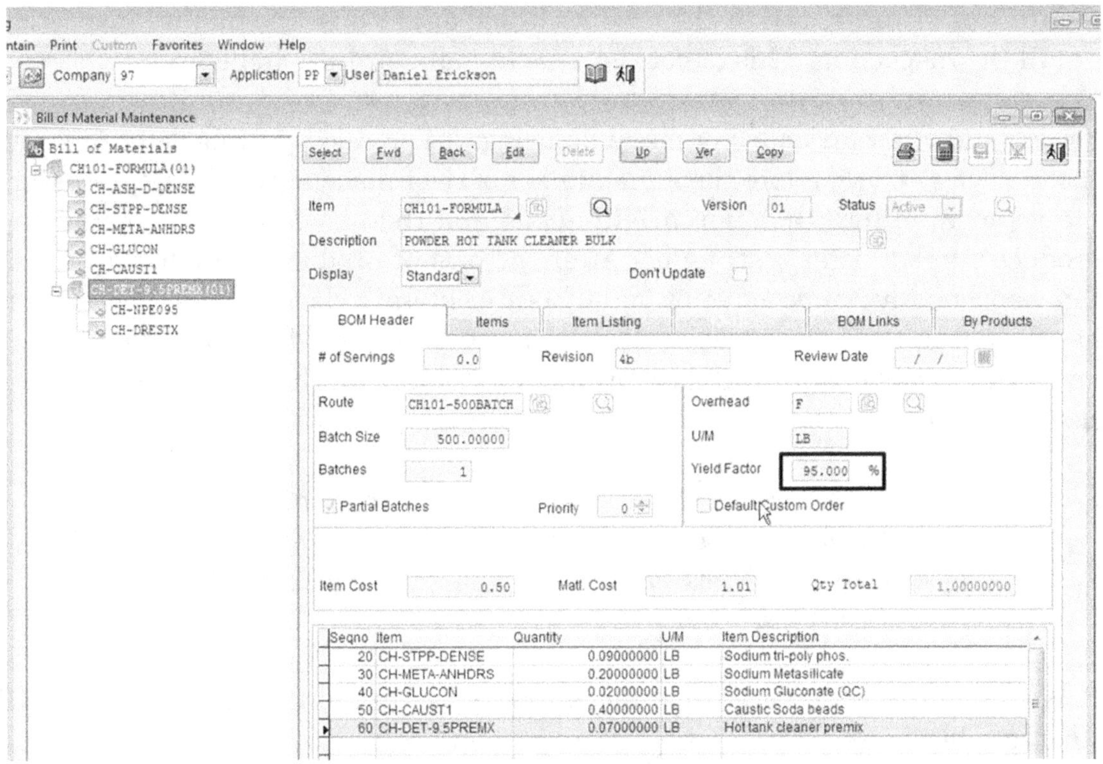

ProcessPro uses the term BOM because the company finds it to be more universally understood. However, ProcessPro has both the ability to have fractional input products on the BOM/recipe/formula as well as have multiple output products and incorporate instructions at a header and line level. Therefore it has true recipe/formula management capabilities. In the screen shot you can see the CH101-Formula has the listed input products. Also notice that the formula has a yield factor of 95 percent. This can be adjusted to any value—and copies of formulas that have slight changes in the manufacturing process or in the input material can be given different yields to reflect reality.

Assigning Completed Production Prior to QA

As most ERP systems are designed for discrete manufacturing, they will often have difficulty in performing a goods receipt from production into stock/inventory **before the item number is assigned.** And this is a problem because in the process industry, an item number will often not be assigned until testing is performed. In discrete manufacturing ERP systems there is a quality hold—but the product either passes or does not pass the hold—however, the item number

would not change, rather the product would either pass the quality inspection or fail the inspection. This is not the case in many process industries.

For instance, in semiconductor manufacturing, the product's testing determines its speed rating and therefore the item number of the finished good. Microprocessors are segmented into the appropriate queue and then assigned the appropriate product number after testing, not after the output product is produced. Before this happens, the semiconductor manufacturer literally does not yet know what it has. However, these companies will still want to record production being complete and some stock being placed into inventory, even though the exact composition of the stock is unknown until testing is complete.

> *"A process ERP system allows the receipt of the production, followed up at a later date by the quality control department to determine and specify quality attributes such as % fat, % solids, brix, acidity, grade, clarity, viscosity, taste, etc."*[104]

Process industry manufacturing companies find several problems with most ERP functionality in regard first to how and when goods receipt can be performed upon untested material, but secondly, how the materials are managed by the system once goods receipt has been performed. Companies can find themselves having to create virtual warehouses to hide the ungraded inventory from MRP. This is necessitated because when the goods receipt has been performed, MRP considers this usable inventory—again an application of a discrete manufacturing requirement.

Process Industry Manufacturing Costing

The specific requirements for process industry manufacturing means having the following costing functionalities:

1. *Net Realizable Value (NRV):* This is costing for supporting primary output, co-product, and bi-product costing. This is the selling price of the inventory minus the selling costs for completion and disposal.[105]

---

[104]Process Industry ERP Requirements
[105]Wikipedia

2. *Multiple Process Definitions:* This is to produce "*the same item that can support planning, scheduling and weighted average costing.*"[106]

3. *Direct Assignment of Usage of Non-Material Inputs:* This includes things like electricity, water and steam. For example, many process industry-manufacturing environments are major electricity consumers—with petroleum refining leading the pack. Electricity is a main component of the overall costs—an external planning system will do the best job of allowing the planner to optimize the production process so that the lowest energy input is required, and of course the ERP system must be able to allocate these costs appropriated rather than simply assigning them to overhead—which is the standard approach to production costing in discrete manufacturing.

4. *Non-Linear Resource Costing*: As was described in Chapter 9: "Constraint Planning and Process Control," process industry resources will often be calculated on the basis of non-linearity. For example a 1,000 pound mixer will not mix 500 pounds in half the time as if 1,000 pounds are loaded—and will cost either the same, or just about the same cost to run. This non-linear consumption applies to both the planning system as well as to costing, which is how this relates to an ERP system, even if an external planning system is used. If the mixer costs $500 to run and 1,000 pounds are mixed, then the cost allocation is 50 cents per pound. However, if 500 pounds are run, then the cost allocation **must be shown** to increase to $1 per pound. In terms of how this is managed by the planning system, it greatly depends upon the particular solution used. PlanetTogether has the best approach to accounting for costs that we (SCM Focus) have seen in a production planning and scheduling vendor. This allows for costs to drive the optimized planning output, although costs are not the objective function. The costs can be given a weight, which will adjust the influence of the costs on the solution. The costs of mixing will be traded off against other criteria—also configurable—while the objective function remains to minimize the duration of the overall Manufacturing Order. This book is not focused on planning methods, but this topic is covered in detail in the SCM Focus Press book, *SuperPlant: Creating a Nimble Manufacturing Enterprise with Adaptive Planning Software.*

[106]Process Industry ERP Requirements

5. *Co-Products and By-Products*: In Chapter 4: “Co-Products and By-Products” we covered how these two output product categories need to be planned. However, the ERP system needs to be able to allocate costing appropriately. For instance while both the primary product and the co-product receive costs, by-products may not have costs allocated to them because they may not be sold product. If they are not sold, by-products are never received into inventory.

ProcessPro allows a variety of methods of product, co-product, and by-product costing. This screen shot shows the very flexible allocation of costs to things ranging from material to variable overhead.

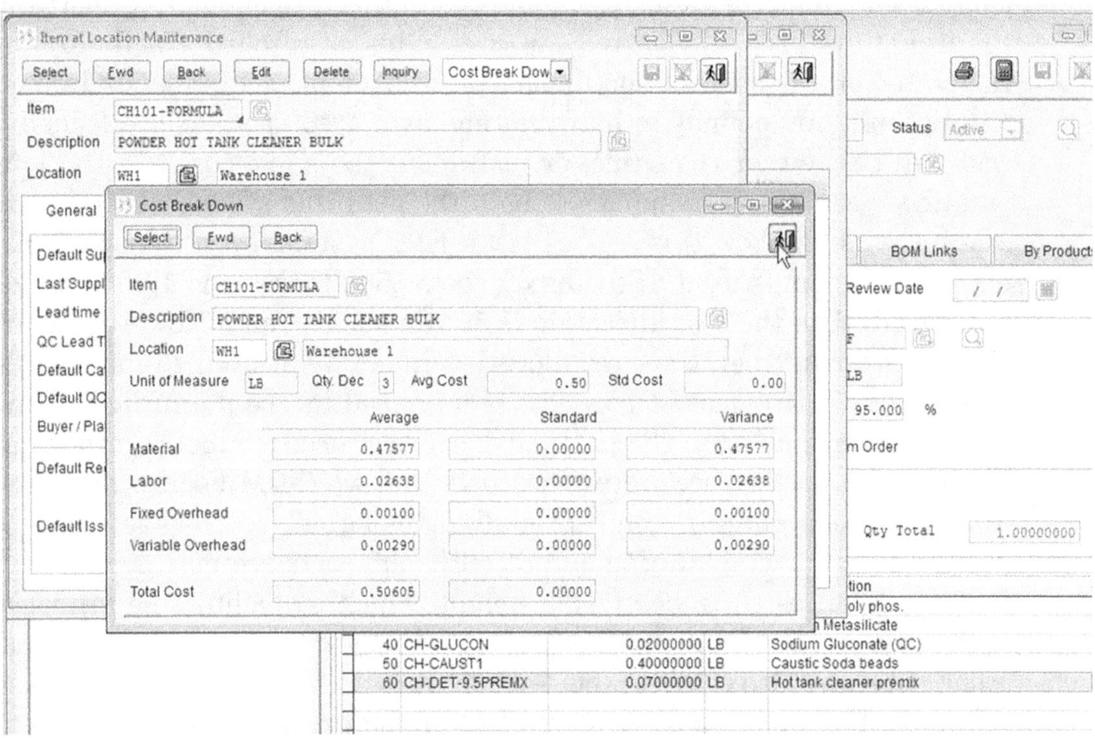

Costing is far more complicated in process industry manufacturing than in discrete and repetitive, and this functionality allows for this allocation per formula.

Date 10/31/13
Time 14:30:59

Page 1

ProcessPro Premier
Costed BOM
Using Average Cost
The Status of this version of the BOM is ACTIVE.

BOM: CH101-FORMULA POWDER HOT TANK CLEANER BULK

Build Quantity: 2500.000 Version: 01 Revision: 4b

Overhead Code: F Route: CH101-500BATCH FG Location: WH1 Raw Location: WH1

Item Description	Ver	Yield/ Scrap	Need Quantity	Unit Cost	Material Cost	Labor Cost	Fixed OH Cost	Variable OH Cost	Extended Cost
CH101-FORMULA	01	95.000	2500.00000000	1.06808	2660.45000		2.50000	7.25000	2670.20000
POWDER HOT TANK CLEANER BULK									
-CH-ASH-D-DENSE	01		579.94736842	1.08250	626.71053				626.71053
Soda Ash, dense									
-CH-STPP-DENSE	01		236.84210526	1.08250	256.38158				256.38158
Sodium tri-poly phos.									
-CH-META-ANHDRS	01		526.31578947	1.08250	569.73684				569.73684
Sodium Metasilicate									
-CH-GLUCON	01		52.63157895	0.75490	39.73159				39.73159
Sodium Gluconate (QC)									
-CH-CAUST1	01		1052.63157895	1.08250	1139.47368				1139.47368
Caustic Soda beads									
-CH-DET-9.5PREMX	01	100.000	184.21052632	0.15426	28.41632				28.41632
Hot tank cleaner premix									
--CH-NPE095	01		105.00000000	0.27063	28.41615				28.41615
T Det N9.5									
--CH-DRESTX	01		79.21052632	0.00000					0.00000
Caustic Soda beads									
Bill of Material Totals:					2660.45000		2.50000	7.25000	2670.20000
							Unit Cost:		1.06808

Reports can be run that show the costing summary for a BOM/Recipe/Formula.

*Adjustment for Feedstock[107]

In both Chapter 2: "Continuous Versus Batch Processing," Chapter 7: "Mixing/Blending/Pooling Operations," and Chapter 9: "Constraint Planning and Process Control" the variability in the quality of feedstock was discussed. Some of those with expertise in the area will list feedstock adjustment/optimization as a requirement for the ERP system (and earlier we saw how this is a claimed capability by an MES software vendor). However, this is only the case if the company has not implemented an external planning system. It is **most desirable** for this variability to be planned by the external planning system. With the right planning system the planning run can be simulated, and the appropriate feedstock can be selected that will provide the desired output product given certain production

[107]*Items with an asterisk are not functionalities that I agree needs to be part of the ERP system, but I include them here because others sometimes include them as requirements for ERP systems. It is very difficult to perform these types of simulations with MRP, which is the only supply and production planning method available in ERP systems. ERP systems have a difficult enough time meeting all of the requirements in this section, and it makes sense to offload those that are planning intensive to an external planning system specialized for process industry manufacturing.

settings. However, if an external planning system is not used, then the ERP system must be able to perform this type of adjustment. But it is extremely difficult to see, given the planning methods available in ERP systems, how this is a suitable application for this problem.

*Master Scheduling Output

This is the ability to translate forecasted demand into a stated master production schedule even when one process results in the production of two or more forecasted items. Master scheduling is also far most effectively performed in the external planning system.

CHAPTER 12

Conclusion

Process industry manufacturing is divided into batch and continuous production, and these categories correlate to discrete and repetitive in terms of their constancy of production. The differentiation is whether the finished good can be disassembled after production into its component parts. Unlike discrete or repetitive manufacturing, process-manufacturing industries have associated output that is produced as a natural consequence of the production process to produce the intended finished good. Co-products are valuable outputs to the manufacturing process and they must be co-planned along with the primary output product. By-products are unwanted outputs of the manufacturing process.

There are a number of manufacturing environments, and each must be managed with supply and production planning approaches that fit those specific environments. The first step toward choosing the right software for the various process industry environments is to understand the differences, and then use this information to develop strategies that match the needs of each environment. Most software is designed around discrete manufacturing, and most books and consultants also speak from the perspective of discrete manufacturing.

Manufacturing accounting, Manufacturing planning, manufacturing execution scheduling, and process control encompasses a large number of problems, and within this continuum there are solutions that are better fits than others. These areas are not a "one-size-fits-all." Because so few consulting companies focus on process industry manufacturing or treat it differently from discrete, it is difficult for companies to get quality outside support on software selection. Manufacturing software, be it ERP, planning, MES or process control tend to be targeted toward discrete manufacturing, often leaving process industry manufacturing companies without software that is designed for their specific requirements.

Process industry manufacturing companies face the same challenges with respect to managing the product information as discrete and repetitive manufacturing companies when it comes to recipe and formula variant management. Unlike SAP, the software vendors PlanetTogether and Demand Works do not use the process industry manufacturing nomenclature in this area, but both are capable of process industry planning because they have designed their functionality to do so. The ability to manage BOM/recipe information in a way congruent with process industry manufacturing needs is a main component of what makes any application process industry capable. While it is certainly true that process-manufacturing recipes can have a great deal of variability, how this is managed is a function of the capability of the application, not what the BOM is called.

Of all of the different types of production environments, process industry is the most sensitive to timing. While all manufacturing categories want to control the time element so as to maximize productivity, process manufacturing is the most time sensitive because time actually changes the characteristics of the input products; work in process, and output product. A second major control over process manufacturing is temperature. For instance, paper, chemicals, metals, power generation, food, beverages, glass and pharmaceuticals are all examples of process industries that use temperature intensively in the manufacturing process.

"Mixing" is the combining of different input products in order to produce a uniform output product. Mixing has a wide range of purposes, one being to provide products on a wide range of specifications. For instance, at a plant that makes industrial lubricants there may be hundreds of lubricant specifications. The plant meets these specifications through mixing/blending. Mixing durations can span from seconds to hours. Mixing is one of the most important quality steps in manufacturing operations that include mixing. Often the quality and consistency of the mix determines a good part of the quality of the finished good. Mixing operations can be simple or complex.

In manufacturing, generally it is common to see the terms "setup" and "changeover" used interchangeably. In process industry manufacturing, a big part of the changeover process is the cleanup or cleaning. The main point of cleanup is to both enable the next processing operation to operate smoothly and to prevent the previously produced item from polluting the next product to be produced. These are different costs and changeover times depending upon the first and second product. This is necessary for proper production scheduling because they reflect the reality of the production floor. Changeovers are typically modeled in planning applications using a changeover table. This master data is then fed to the optimizer of a planning system. Changeover planning can be performed in a variety of ways, with one of the best that we have seen being by attributes.

Manufacturing execution scheduling (MES) software covers the phase, after planning and works in conjunction with the process control system(s). MES systems take up the functionality space between the ERP system and the process control systems (which are discussed in Chapter 9: "Constraint Planning and Process Control." MES systems tend to be applied most frequently in the process industry manufacturing sub-categories where tracking is very important for regulatory compliance.

When it comes to ERP, there are specific areas of functionality required that discreet and repetitive manufacturing simply does not use. Some ERP vendors say they have this functionality, but most ERP systems leave process industry manufacturing companies wanting more. This book highlighted one software vendor that bases its application and business on serving the process industry manufacturing market exclusively.

The focus of this book has been to document and explain the specific requirements that are required by process industry manufacturing software. Secondly, specific software vendors were used to demonstrate that most of this functionality does exist in commercial applications. From this I hope that the book was able to demonstrate that there really is no reason for so many process industry manufacturers to accept enterprise software that is not specifically designed for their needs.

Other Books from SCM Focus

Bill of Materials in Excel, ERP, Planning and PLM/BMMS Software

http://www.scmfocus.com/scmfocuspress/the-software-approaches-for-improving-your-bill-of-materials-book/

Constrained Supply and Production Planning in SAP APO

http://www.scmfocus.com/scmfocuspress/select-a-book/constrained-supply-and-production-planning-in-sap-apo/

Enterprise Software Risk: Controlling the Main Risk Factors on IT Projects

http://www.scmfocus.com/scmfocuspress/it-decision-making-books/enterprise-software-project-risk-management/

Enterprise Software Selection: How to Pinpoint the Perfect Software Solution using Multiple Information Sources

http://www.scmfocus.com/scmfocuspress/it-decision-making-books/enterprise-software-selection/

Enterprise Software TCO: Calculating and Using Total Cost of Ownership for Decision Making

http://www.scmfocus.com/scmfocuspress/it-decision-making-books/enterprise-software-tco/

Gartner and the Magic Quadrant: A Guide for Buyers, Vendors and Investors

http://www.scmfocus.com/scmfocuspress/it-decision-making-books/gartner-and-the-magic-quadrant/

Inventory Optimization and Multi-Echelon Planning Software

http://www.scmfocus.com/scmfocuspress/supply-books/the-inventory-optimization-and-multi-echelon-software-book/

Multi-Method Supply Planning in SAP APO

http://www.scmfocus.com/scmfocuspress/select-a-book/multi-method-supply-planning-in-sap-apo/

Planning Horizons, Calendars and Timings in SAP APO

http://www.scmfocus.com/scmfocuspress/select-a-book/planning-horizons-calendars-and-timings-in-sap-apo/

Replacing Big ERP: Breaking the Big ERP Habit with Best of Breed Applications at a Fraction of the Cost

http://www.scmfocus.com/scmfocuspress/erp-books/replacing-erp/

Setting up the Supply Network in SAP APO

http://www.scmfocus.com/scmfocuspress/select-a-book/setting-up-the-supply-network-in-sap-apo/

SuperPlant: Creating a Nimble Manufacturing Enterprise with Adaptive Planning Software

http://www.scmfocus.com/scmfocuspress/production-books/the-superplant-concept/

Supply Chain Forecasting Software

http://www.scmfocus.com/scmfocuspress/the-statistical-and-consensus-supply-chain-forecasting-software-book/

Supply Planning with MRP, DRP and APS Software

http://www.scmfocus.com/scmfocuspress/supply-books/the-supply-planning-with-mrpdrp-and-aps-software-book/

The Real Story Behind ERP: Separating Fiction from Reality

http://www.scmfocus.com/scmfocuspress/erp-books/the-real-story-behind-erp/

Spreading the Word

SCM Focus Press is a small publisher. However, we pride ourselves on publishing the unvarnished truth, which most other publishers will not publish. If you felt like you learned something valuable from reading this book, please spread the word by adding a review to our page on Amazon.com.

APPENDIX A

AspenTech Training

It's no secret that companies that implement enterprise software tend to use only a small portion of the overall functionality that a vendor offers. In some cases this functionality can be sufficient to provide the necessary return on investment to justify the purchase. However, it is quite common for implementing companies to overlook functionality that could help them get more value from the application.

The common refrain is that training is necessary to realize all the functionality of an application; however, onsite training can be tricky in that such a large amount of information is condensed into a short time period. For instance, when I have taken training courses myself, I often feel it necessary to dedicate an entire week following the course to reviewing the material to which I was exposed, and I'm only really able to do this when I'm not on a project. Most full-time employees do not have this luxury; they must be back and productive the week following their training, usually doing their normal jobs. This is why the instructional material is so important, in that it allows users and consultants to train on their own time and at their convenience.

I have always found well-written product manuals to be extremely helpful. However, for interacting with an application, I don't think anything beats video. Even though this viewpoint is shared by a number of trainers, video training in enterprise software is still relatively uncommon. Therefore, it was nice to find the training that is available at AspenTech. AspenTech offers a number of instructional videos on its website that anyone can watch (customers and noncustomers). However, they also offer even more videos after the software purchase. The screen shot below shows that these videos are combined in a user interface that allows the viewer to choose a topic of interest.

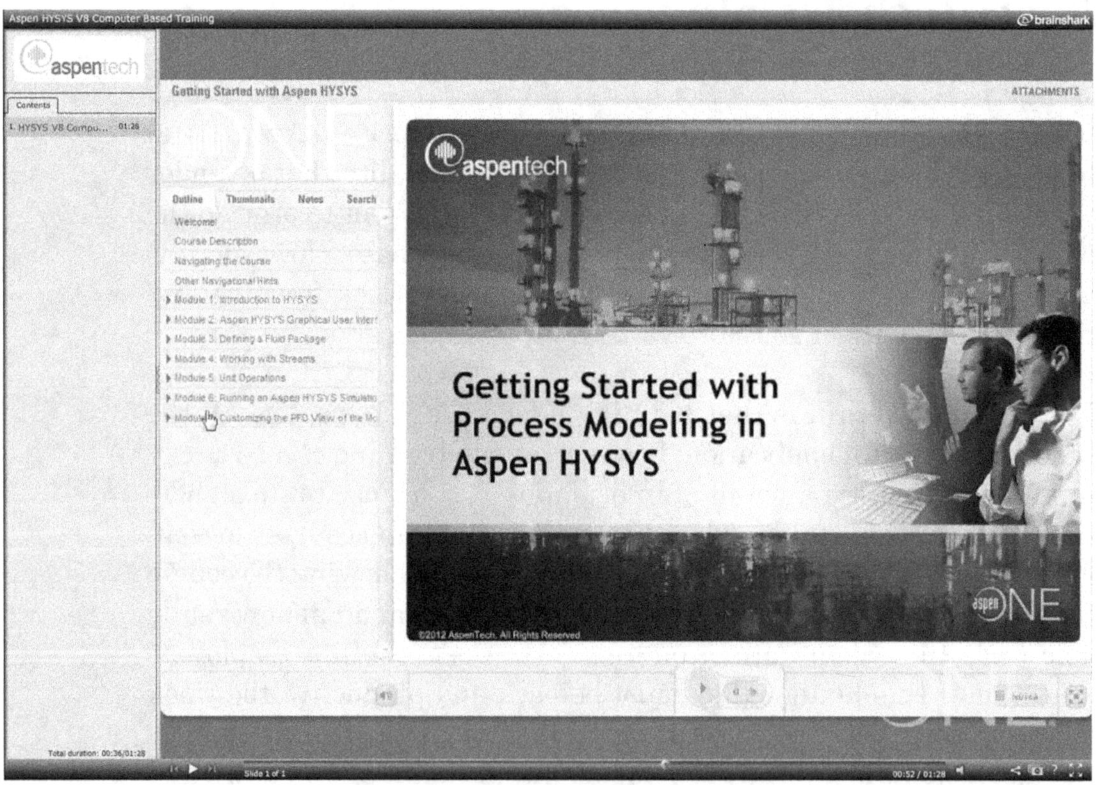

Here the viewer can select from a variety of videos.

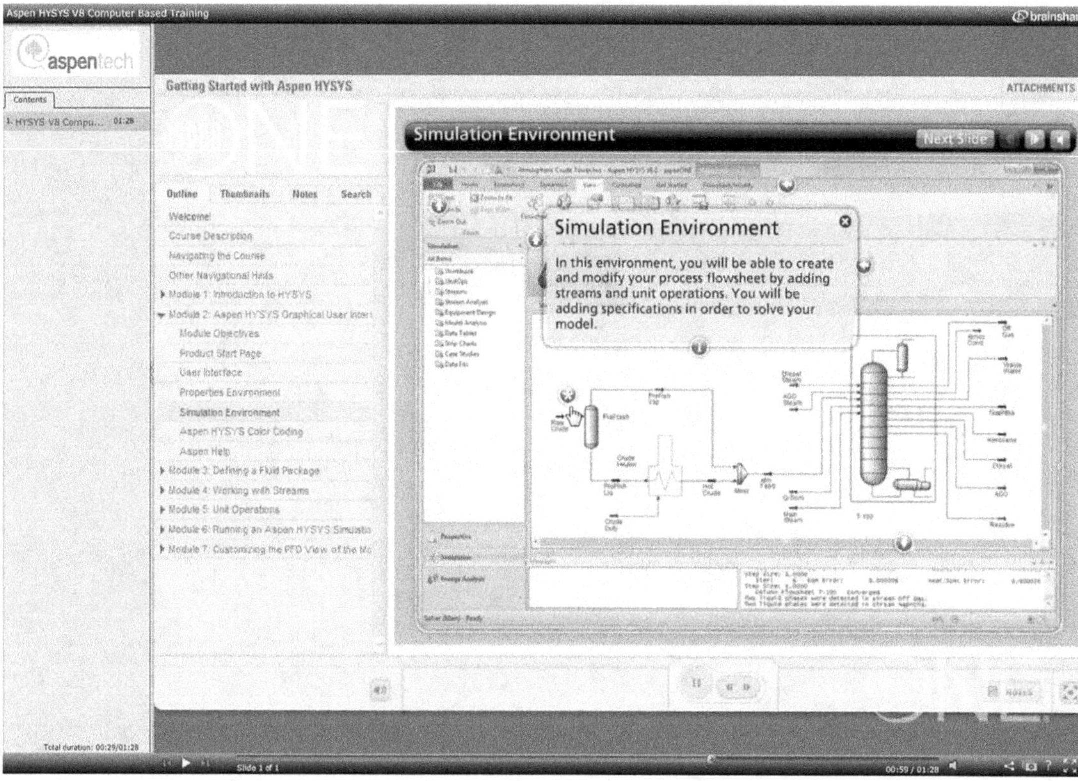

This allows the viewer to hover over various areas of the AspenTech user interface and bring up the details of each area.

I have yet to see training—and in particular content—viewed as a differentiator in software selection; however, it is most important. Training is viewed universally as important, but innovative approaches such as this one displayed by AspenTech, which has really put a lot of thought into how their applications will be learned, and which provides a high degree of convenience to users and customers, should really be evaluated during software selection.

APPENDIX B

AspenTech and Feedstock Projection

What is quite interesting about AspenTech is the upstream use of their planning software. In discrete manufacturing, the raw materials and other input products are mostly stable before production begins. There is often something called a scrap factor, which adjusts for damaged product.

Here's an example of a scrap factor. A stamping machine is known to produce two defects out of three hundred stampings. An extra four units would be ordered when the production order is created in order to ensure that the factory has sufficient stock to complete the production order.

However, process industry planning is quite a bit different in this regard, as the input product, or feedstock, can result in very different output products, as well as different manufacturing lead times. A good example of this is sour crude versus sweet crude. Sweet crude requires less processing in order to generate the output product of gasoline, and more gasoline will result from a gallon of sweet crude than a gallon of sour crude. Therefore, if a plant has a desired output, and has a choice of a range of feedstock, it is quite desirable to be able to simulate the production run first and choose the feedstock after the fact. As one can see, there are multiple dimensions that can be controlled—everything from the feedstock selected to the duration of the manufacturing process (which is related to the energy consumption of the manufacturing process), to the yield of different finished goods materials. All of these dimensions interact; for instance, more processing time can mean the ability to use a lower cost feedstock.

This is a perfect environment in which to computerize the decision making because the user can simply specify the desired outcomes and allow the model to provide the answer. This is a far more useful and sophisticated approach to optimization than using a single dimension such as cost to control the outcome.

APPENDIX C

AspenTech and Exceptions

One could debate if exception management is more important for discrete, repetitive, or process industry manufacturing, but AspenTech's alerts/exceptions have been intelligently designed. To understand why they are so effective, it's necessary to compare them to a weaker alert system, which can be found in SAP APO (shown in the following screen shot).

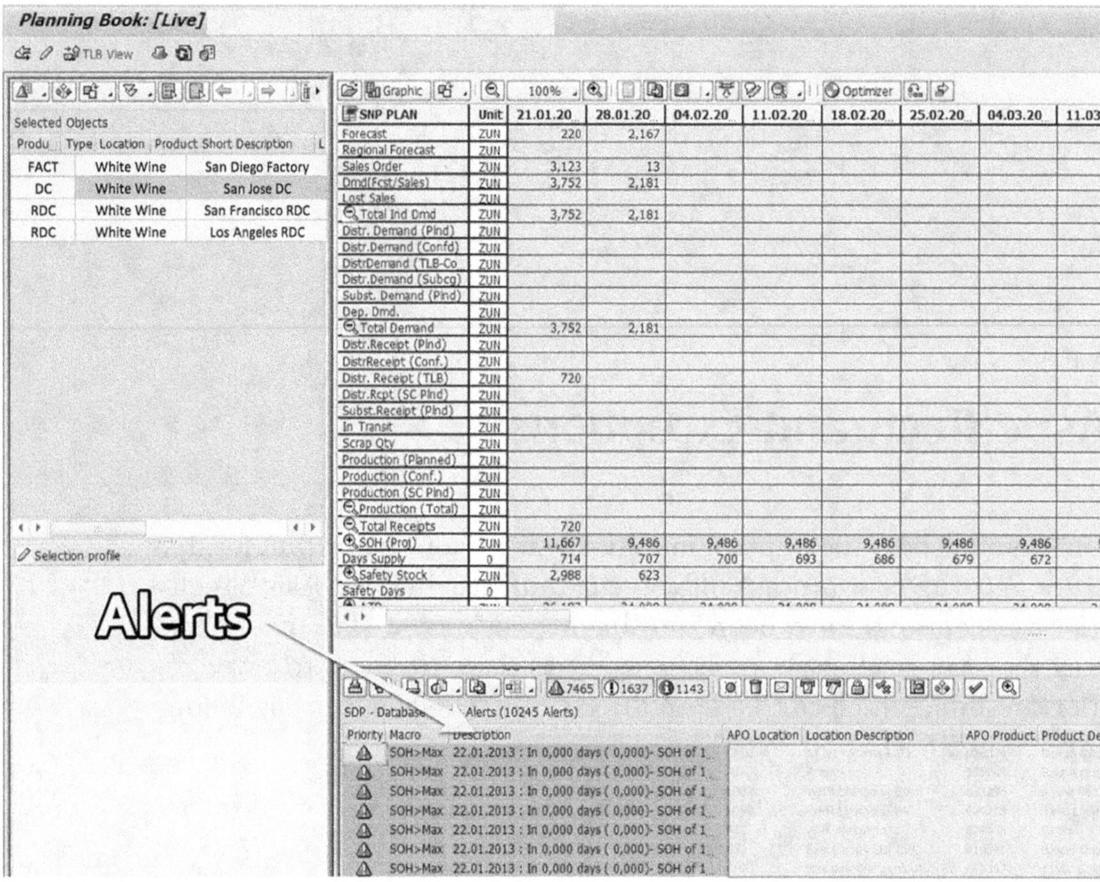

SAP APO has all types of problems with its alert configuration, and the alert screen was actually better when I worked with APO back in 2003. However, they actually degraded the APO alert monitor to the point where it is problematic to work with. This screen shot is a view within an APO Planning Book. I don't want to get into too much detail about the APO Planning Book except to say that it is a main user interface for several of the most popular SAP APO modules. For those interested in more detail, I cover the Planning Book in the following article:

http://www.scmfocus.com/sapplanning/2011/06/09/the-planning-book-key-figure-and-macro-definitions/

Alerts can be viewed in a few places in APO, but one of these places is in the Planning Book. This allows planners to be alerted to issues when they are drilled into a specific selection—not always, but normally a product, a location or a product-location combination. These alerts, along the bottom of the screen, are quite generic. They tell you roughly what is wrong, but they do not really provide very much depth of understanding.

The AspenTech alerts, on the other hand, are quite a bit more descriptive. In fact, AspenTech's alerts are rare in a planning application, and I show an example in the screen shot below:

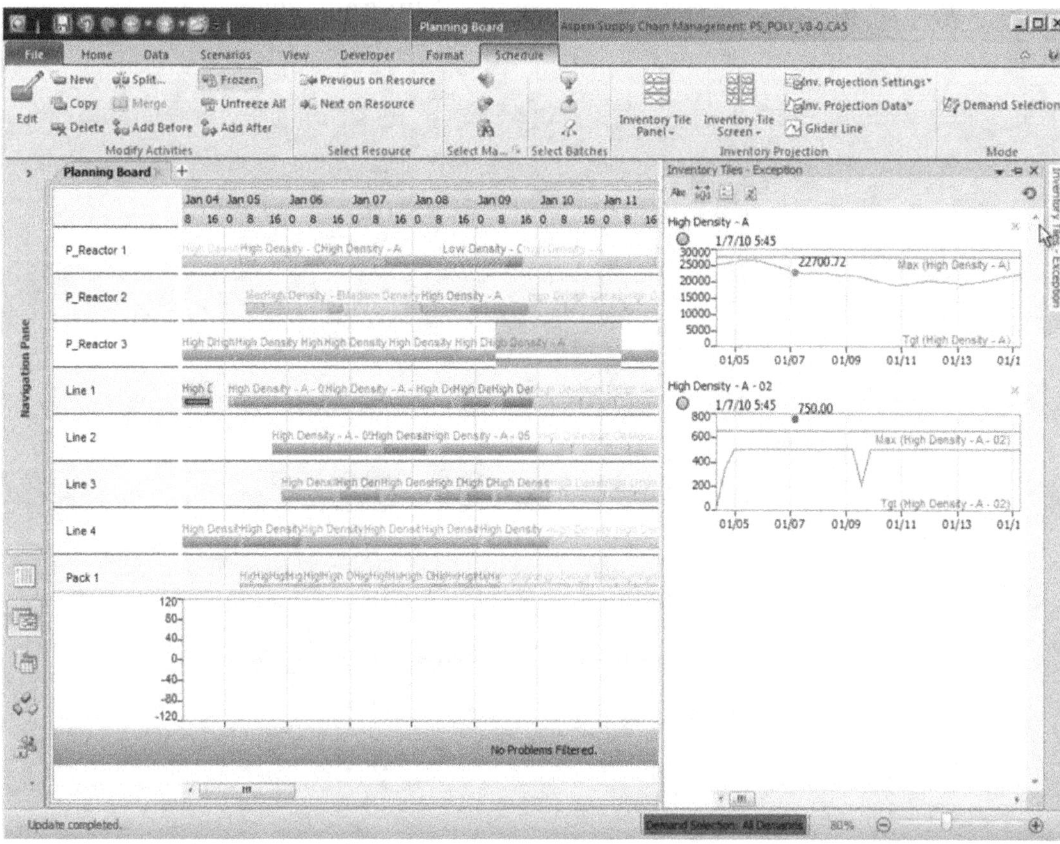

Notice that unlike the SAP APO alerts, the AspenTech alerts carry more information. They show a graph per alert. These graphs compare the Maximum versus the Target density. The application is telling the planner (who recently placed a new planned production order into the schedule) that the plant is scheduled to produce two items that are out of the target density. Notice the Gantt chart to the left shows various planned production orders of various densities. The planner can then make the necessary adjustments.

This particular alert was caused by the addition of a planned production order to an existing schedule. The alert side panel may be opened whenever alerts are to be reviewed and kept closed when the planner is not interested in reviewing them.

The Directness of the Adjustment

One of the major weaknesses of the alerts in APO is that it is time consuming and often indirect to make the changes that will address the alerts. AspenTech on the other hand is quite direct. I was able to immediately know which changes to make and this makes alerts both easier to set up and easier to use.

Links in Book

Chapter 1

http://www.scmfocus.com/writing-rules/

http://www.scmfocus.com

http://www.scmfocus.com/scmfocuspress/production-books/process-industry-planning/

Chapter 2

http://www.scmfocus.com/supplyplanning/2011/10/02/the-four-factors-that-make-up-the-master-production-schedule/

http://www.scmfocus.com/supplyplanning/2012/07/25/the-misuse-and-overuse-of-the-term-mrp/

Blast Furnace

http://www.youtube.com/watch?v=64c2mHdzxf8

Oil Refinery

http://www.youtube.com/watch?v=2xzYf8IL_FE

Loom

http://www.youtube.com/watch?v=V5e1YKkZFhc

Paper Manufacturing

http://www.youtube.com/watch?v=jGH7kQ30SKo

Nuclear Reactor

http://www.youtube.com/watch?v=MGj_aJz7cTs

Steam Turbine

http://www.youtube.com/watch?v=AyAd-gLO9CE

Hydroelectricity

http://www.youtube.com/watch?v=rnPEtwQtmGQ

Water Treatment

http://www.youtube.com/watch?v=SC8QRs77iH0

Steel Rolling

http://www.youtube.com/watch?v=6xnKmt_gsLs

Fractional Distillation

http://www.youtube.com/watch?v=Z6OyNB8V7Hc

Cow Milking

http://www.youtube.com/watch?v=cPLTUnOOAN0

How Paint is Made

http://www.youtube.com/watch?v=hX_qTE5xlb0

http://www.kodak.com/ek/US/en/Industrial_Materials_Group_LHN/Specialty_Chemicals/Products.htm

Chapter 3

http://help.sap.com/saphelp_SCM700_ehp01/helpdata/en/48/aae32a8740356ce10000000a421937/frameset.htm

Chapter 5

http://www.scmfocus.com/productionplanningandscheduling/2011/11/10/the-problem-with-excessive-costing-bills-of-material/

http://www.softwaredecisions.org

http://www.hamilton-grant.com/The-HG-Solution/HG-Recipe-Management/Formula-Optimizer

Chapter 8

http://www.scmfocus.com/supplyplanning/2011/07/09/what-is-your-supply-planning-optimizer-optimizing/

http://www.scmfocus.com/supplyplanning/2011/07/24/why-the-cost-of-missed-demand-must-always-be-set-so-much-higher-than-the-profit-margin/

http://www.scmfocus.com/scmhistory/2010/07/the-rise-and-decline-for-supply-chain-optimization/

http://www.scmfocus.com/supplychainsimulation/

http://www.scmfocus.com/supplyplanning/2011/07/10/customizing-the-optimization-per-supply-chain-domain/

http://www.scmfocus.com/productionplanningandscheduling/2011/01/20/cost-vs-duration-based-optimization-for-scheduling/

Appendix C

http://www.scmfocus.com/sapplanning/2011/06/09/the-planning-book-key-figure-and-macro-definitions/

References

Amendola, Joseph. Rees, Nicole. *Understanding Baking: The Art and Science of Baking.* Wiley. 2002.

American Iron and Steel Institute. *How a Blast Furnace Works.* http://www.steel.org/Making%20Steel/How%20Its%20Made/Processes/How%20A%20Blast%20Furnace%20Works%20larry%20says%20to%20delete.aspx.

AspenONE Petroleum Supply Chain. http://www.aspentech.com/products/aspenONE-Petroleum-Supply-Chain/.

AspenOne Supply Chain for Chemicals. AspenTech. 2011.

Backflushing. Accessed October 19, 2013. http://en.wikipedia.org/wiki/Batch_production.

Bartholomew, Doug. Realizing ERP's Untapped Potential. Pharmacy Manufacturing. Sept 7, 2005.

Batch Production. Accessed June 12, 2013. http://en.wikipedia.org/wiki/Backflush_accounting

Benson, Ross. An MES Reality Check. Pharmaceutical Manufacturing. 2005. http://www.pharmamanufacturing.com/articles/2005/322/

Berk, Zeki. *Food Process Engineering and Technology.* Academic Press, 2008.

Bernstein, William J. *Masters of the Word: How Media Shaped History From the Alphabet to the Internet.* Grove Press. 2013

Blast Furnace. Accessed June 21, 2013. http://en.wikipedia.org/wiki/Blast_furnace.

Bordeleau, Stephane. *Where Canada's Surplus Energy Goes*, CBCNews, March 30, 2011. http://www.cbc.ca/news/canada/where-canada-s-surplus-energy-goes-1.1109321

Brewing. Accessed October 20, 2013. http://en.wikipedia.org/wiki/Brewing.

By-product. Accessed October 19, 2013. http://en.wikipedia.org/wiki/By-product.

Catch Weight Software. MIBAR. http://www.mibar.net/Industries/Food-and-Beverage/Catch-Weight-Software.aspx.

Cheese. Accessed October 25, 2013. http://en.wikipedia.org/wiki/Cheese.

Chemical Industry. Accessed October 14, 2013. http://en.wikipedia.org/wiki/Chemical_industry.

Coca-Cola Formula. Access June 26, 2013. http://en.wikipedia.org/wiki/Coca-Cola_formula.

Commodity Chemicals. Accessed October 13, 2013. http://en.wikipedia.org/wiki/Commodity_chemicals.

Continuous Production. Accessed March 7, 2013. http://en.wikipedia.org/wiki/Continuous_production.

Co-product. Accessed October 18, 2013. http://en.wikipedia.org/wiki/Coproduct

Czochralski Process. Accessed May 1, 2013. http://en.wikipedia.org/wiki/Czochralski_process.

Dairy Farming. Accessed October 20, 2013. http://en.wikipedia.org/wiki/Milk_production.

Dessoukey, Maged, Brian Kijowski, and Sushil Verma. *Simultaneous Batch and Scheduling for Chemical Processing with Earliness and Tardiness Penalties.* Production and Operations Management, Vol. 8, No. 4 (1999).

Distilled Beverages. Accessed October 20, 2013. http://en.wikipedia.org/wiki/Distilled_beverage.

Distillation. Accessed October 20, 2013. http://en.wikipedia.org/wiki/Distillation.

Doles, David. *What is Specialty Chemical Manufacturing?* Society of Chemical Manufacturers and Affiliates. http://www.socma.com/specialtymanufacturing.html.

El-Berishy, Nagham. *An Advanced Production Planning and Scheduling System: For Batch Process Industry*. LAP LAMBERT Academic Publishing, 2011.

Engine Efficiency. Accessed July 5, 2013. https://en.wikipedia.org/wiki/Engine_efficiency.

Epp, Dianne N. *Chemical Manufacturing: The Process of Mixing*. Terrific Science Press, 2000.

External Combustion Engine. Accessed June 13, 2013. http://en.wikipedia.org/wiki/External_combustion_engine.

Floyd, Raymond C. *Liquid Lean: Developing Lean Culture in the Process Industry*. Productivity Press, 2010.

Food and Beverage Distribution & Manufacturing Software. MIBAR. http://www.mibar.net/Industries/Food-and-Beverage.aspx.

Formulas and Recipes. http://www.tgiltd.com/erp-software/formulas-and-recipes-software.html.

Fluid Catalytic Cracking. Accessed June 19, 2013. http://en.wikipedia.org/wiki/Fluid_catalytic_cracking.

Geipel-Kern, Anke. *10 Trends for the Process Industry You Should Know*. Process Worldwide. June 6, 2012. http://www.process-worldwide.com/management/markets_industries/articles/366973/.

Global Specialty Chemicals. MarketLine Industry Profile. May 2012. http://2012e.igem.org/wiki/images/f/fb/Global_Specialty_Chemicals_Profile.pdf.

Gunther, Hans-Otto. *Advanced Planning and Scheduling Solutions in Process Industry*. GOR Publications, 2009.

Gupte, Askay, Shabbir Ahmed, Santanu S. Dey, and Myun Seok Cheon. *Pooling Problem*. Optimization Online, 2012.

He, Yaohua. *Meta-Heuristics for Large Scale Process Scheduling*. VDM Verlag Dr. Müller, 2009.

History of Wine. Accessed October 21, 2013. http://en.wikipedia.org/wiki/History_of_wine.

How to Choose ERP Software for Process Manufacturers. ProcessPro.

Impact of Formula-Based ERP Applications on Pharmaceutical Manufacturers. CDC Software. http://www.pharmamanufacturing.com/assets/wp_downloads/pdf/CDC_ERP.pdf.

Joint Product Costing http://help.sap.com/saphelp_46c/helpdata/en/7e/cb854743a311d189ee0000e81ddfac/frameset.htm.

Kallrath, Josef . *Planning and scheduling in the process industry.* OR Spectrum, Vol. 24, Issue 3 (2002): 219-250. http://link.springer.com/article/10.1007/s00291-002-0101-7.

Kallrath, Josef, and Thomas I. Maindl. *Real Optimization with SAP APO.* Springer Press, 2006.

Katz, Sandor Ellix. *The Art of Fermentation: An In Depth Exploration of Essential Concepts and Processes from Around the World.* Chelsea Green Publishing, 2012.

King, Peter L. *Lean for the Process Industries: Dealing with Complexity.* Productivity Press, 2009.

Koch, Christopher, Derek Slater, and E. Baatz. *The ABCs of ERP.* http://paginas.fe.up.pt/~mgi00011/ERP/abcs_of_erp.htm.

Kvaalen, Eric. Wankat, Philip C. McKenzie, *Bruce A. Alcohol Distillation: Basic Principles, Equipment, Performance, Relationships and Safety.* Purdue University. http://www.extension.purdue.edu/extmedia/AE/AE-117.html.

Leffler, William L. *Petroleum Refining in Nontechnical Language (4th Edition).* PennWell Corporation, 2008.

Loom. Accessed May 20, 2013. http://en.wikipedia.org/wiki/Loom.

Malting. Accessed October 23, 2013. http://en.wikipedia.org/wiki/Malting#Malting.

Manufacturing Execution Scheduling. LinkedIn. Accessed October 19, 2013. http://www.linkedin.com/groups/MES-process-industry-vs-discrete-67917.S.163781745.

Magdassi, Shlomo. *The Chemistry of Inkjet Inks.* World Scientific. July 2009.

Mixing (Process Engineering). Accessed June 20, 2013. http://en.wikipedia.org/wiki/Mixing_(process_engineering).

Mill. Accessed October 22, 2013. http://en.wikipedia.org/wiki/Mill_(grinding).

Mokhatab, Saeid. Poe, A. William. *Handbook of Natural Gas Transmission and Processing, Second Edition.* Gulf Professional Publishing. 2012.

Muller, D. de Villiers, P.G.R. Humphries, G. *A Holistic Approach to Control and Optimisation of an Industrial Crushing Circuit,* AspenTech.

Murray, Paul. "What's Your Game Plan for Implementing MES?" Pharmacy Manufacturing. 2006. http://www.pharmamanufacturing.com/articles/2006/173/.

National Collection of Yeast. Accessed October 22, 2013. http://en.wikipedia.org/wiki/National_Collection_of_Yeast_Cultures.

Natural Gas. Accessed October 21, 2013. http://en.wikipedia.org/wiki/Natural_gas.

Net Realizable Value. Accessed on October 19, 2013. http://en.wikipedia.org/wiki/Net_realizable_value.

Nienow, A.W. and M.F. Edwards. *Mixing in the Process Industries.* Plant a Tree Press, 2001.

Oldest Winepress http://www.bbc.co.uk/news/world-europe-12158341.

Ophardt, Charles E. Visual Chembook. Elmhurst College. 2003.

Control of (the) Fermentation Process. Amtech Process Instruments. 2003.

Paper. Accessed June 13, 2013. https://en.wikipedia.org/wiki/Paper.

Paper Machine. Accessed June 25, 2013. http://en.wikipedia.org/wiki/Paper_machine.

Peak Oil. Accessed August 6, 2013. http://en.wikipedia.org/wiki/Peak_oil.

Petroleum Coke. Accessed July 31, 2013. https://en.wikipedia.org/wiki/Petroleum_coke.

Petroleum Refining, Emissions Factors, 1-1995. https://docs.google.com/viewer?url=http%3A%2F%2Fwww.epa.gov%2Fttnchie1%2Fap42%2Fch05%2Ffinal%2Fc05s01.pdf.

Planning & Scheduling for Batch Process Industries. AspenTech. 2009.

Printing Press. Accessed October 15, 2013. http://en.wikipedia.org/wiki/Printing_press.

Process Control. Accessed May 29, 2013. http://en.wikipedia.org/wiki/Process_control.

Process ERP Partners LLC. *What Makes Process Process?* South Africa Production & Inventory Control Society, 1998.

Process Industry Manufacturing Software: ERP, Planning, Process Control, Recipe Mgmt. & MES Software.

Process Manufacturing. Accessed May 22, 2013. http://en.wikipedia.org/wiki/Process_manufacturing.

Process Manufacturing. http://help.sap.com/saphelp_46c/helpdata/en/8a/1a1f674e4211d182be0000e829fbfe/frameset.htm.

Production Planning – Process Industries (PP-PI). http://help.sap.com/saphelp_470/helpdata/en/89/a419ff461e11d182b50000e829fbfe/content.htm.

Process Industry ERP Requirements. Wonderware. 2000.

RAND Corporation. *Standards for High-Quality Research and Analysis.* http://www.rand.org/standards.html.

Recycle and Bypass Processes. http://facstaff.cbu.edu/rprice/lectures/recycle.html.

Rambalee, P. Gous, G. de Villiers, P.G.R, McCulloch, N. Humphries, G. *Control and Stabilization of a Multiple Boiler Plant: An APC Approach.* AspenTech.

Roessler, Dan. *Enabling Process Manufacturers to Leverage Resource Expertise Globally.* AspenTech. 2011.

Rokohl, Laura. "Visibility and Integration – The Key Ingredients for a Successful Supply Chain." AspenTech. 2012.

Rolling (metalworking). Accessed October 12, 2013. http://en.wikipedia.org/wiki/Rolling_(metalworking).

Savu, Andreea. Lazea A, Gheorghe. Agachib, Paul-Şerban. *Optimization and Advanced Control for Thermal Cracking Processes.* 20th European Symposium on Computer Aided Process Engineering. 2012.

Scholten, Bianca. *MES Guide for Executives: Why and How to Select, Implement, and Maintain a Manufacturing Execution System.* International Society of Automation. 2009.

Seborg, Dale E. Mellichamp, Dale A. *Process Dynamics and Control 2nd Edition.* Wiley. 2003.

Sectors of Chemical Industry. TechnoFunc. http://www.technofunc.com/index.php/domain-knowledge/chemicals-industry/item/sectors-of-chemical-industry.

Semiconductor Device Fabrication. Accessed July 3, 2013. http://en.wikipedia.org/wiki/Semiconductor_device_fabrication.

Sewage Treatment. Accessed October 21, 2013. http://en.wikipedia.org/wiki/Sewage_treatment.

Singh, Harpal. *Supply Chain Planning for the Process Industry.* Supply Chain Consultants, 2010.

Snapp, Shaun. *Enterprise Software Selection: How to Pinpoint the Perfect Software Solution Using Multiple Information Sources.* SCM Focus Press, 2013.

Snapp, Shaun. *Multi-method Supply Planning in SAP APO.* SCM Focus Press, 2013.

Snapp, Shaun. *Superplant: Creating a Nimble Manufacturing Enterprise with Adaptive Planning Software.* SCM Focus Press, 2013.

Snapp, Shaun. *Supply Planning with MRP, DRP and APS Software.* SCM Focus Press, 2012.

Snapp, Shaun. *The Bill of Materials in Excel, ERP, Planning and PLM/BMMS Software.* SCM Focus Press, 2013.

Snapp, Shaun. *The Real Story Behind ERP: Separating Fiction from Reality.* SCM Focus Press, 2013.

Sour Crude Oil. Accessed August 7, 2013. http://en.wikipedia.org/wiki/Sour_crude_oil.

Standard products from variable quality ingredients. http://www.hamilton-grant.com/The-HG-Solution/HG-Formula-Optimizer.

Steps in Making Cheese. http://www.eatwisconsincheese.com/wisconsin/how_cheese_is_made.aspx.

Subramaniam, Anand. *Manufacturing Execution System (MES).* 2009.

Suzuki, Tokutaro. *TPM in Process Industries (Step-By-Step Approach to TPM Implementation).* Productivity Press, 1994.

Sulfur. Accessed October 20, 2013. http://en.wikipedia.org/wiki/Sulfur.

Sulfur History, Georgia Gulf Sulfur Corporation. http://www.georgiagulfsulfur.com/history.htm.

Sweet Crude Oil. Accessed June 21, 2013. http://en.wikipedia.org/wiki/Sweet_crude_oil.

Sweetening Up the Crude. Saudi Aramco World, Vol. 11, No. 1 (1960). http://www.saudiaramcoworld.com/issue/196001/sweetening.up.the.crude.htm.

Tekchandaney, Jayesh. *Batch Mixing.* Last modified October 9, 2009. http://www.brighthubengineering.com/manufacturing-technology/52042-all-about-batch-vs-continuous-mixing/.

Tekchandaney, Jayesh. *Introduction to Mixing Technology.* Last modified December 11, 2012. http://www.brighthubengineering.com/manufacturing-technology/41798-introduction-to-mixing-technology/.

Tekchandaney, Jayesh. *Material Properties Affecting Solids Blending and Blender Selection: Bulk Density.* Last modified October 25, 2009. http://www.brighthubengineering.com/manufacturing-technology/53444-bulk-density-solids-blending-and-blender-selection/.

Use of Process Analytics for Cracking Processes in Refineries. Siemens. 2007.

Using PlanetTogether for Effective Scheduling in the Food Process Industry. http://planettogether.com/wp-content/uploads/2011/06/Using-PlanetTogether-for-Effective-Scheduling-in-the-Food-Process-Industry.pdf.

Wailgum, Thomas. *Why ERP Systems Are More Important Than Ever.* January 29, 2008. http://www.cio.com/article/177300/Why_ERP_Systems_Are_More_Important_Than_Ever?page=3&taxonomyId=3000.
http://help.sap.com/saphelp_470/helpdata/en/89/a42c7a461e11d182b50000e829fbfe/frameset.htm
http://www.lifescienceleader.com/magazine/past-issues3/item/3502-is-the-reign-of-batch-processing-coming-to-an-end.

Water Purification. Accessed October 22, 2013. http://en.wikipedia.org/wiki/Water_purification.

Weill, Peter. “The Relationship Between Investment in Information Technology and Firm Performance: A Study of the Valve Manufacturing Sector.” Information Systems Research. Dec 1992.

Winemaking. Accessed October 22, 2013. http://en.wikipedia.org/wiki/Yeast_in_winemaking.

Vendor Acknowledgments and Profiles

I have listed brief profiles of each vendor with screen shots included in this book below.

AspenTech

AspenTech is the largest provider of production planning and scheduling software to the process industry. With integrated AspenONE solutions, process manufacturers can implement best practices for optimizing their engineering, manufacturing, and supply chains. As a result, AspenTech customers are better able to achieve their operational excellence goals: increasing capacity, improving margins, reducing costs, and becoming more energy efficient.

www.aspentech.com

Demand Works

Demand Works is a best-of-breed demand and supply planning vendor that emphasizes flexible and easy-to-configure solutions. This book only focuses on the supply planning functionality within their Smoothie product, which includes MRP and DRP.

www.demandworks.com

Hamilton Grant

Hamilton Grant focuses on the food and beverage industry offering recipe management solutions. The depth of functionality of our software is genuinely unique. The list is too numerous to mention. But to give some examples: the abilities to manipulate formulations; to try out "what if?" scenarios; to build production formulations from trial formulations; to easily compare any aspects of a set of trial or production formulations; to easily customise the data sets and build in your own specific calculations; to build in the effect of processes (e.g., drying; extraction of whey; vitamin degradations); to define formats and legislation for the calculation of ingredient declarations.

http://www.hamilton-grant.com/

PlanetTogether

PlanetTogether, software developer of Galaxy APS (Advanced Planning & Scheduling), enables manufacturers to eliminate spreadsheets and connect multi-plant operations for real-time visibility and collaboration. Users report 50 percent reductions in inventory and labor costs, faster time-to-delivery, and a 6-month return on investment. PlanetTogether serves nearly one hundred clients in process and discrete manufacturing in a broad range of vertical segments.

www.planettogether.com

ProcessPro

ProcessPro is designed specifically for manufacturers that combine batches of mixtures, ProcessPro seamlessly integrates all aspects of plant operation, from beginning order entry through manufacturing, packaging, shipping, inventory, and accounting. ProcessPro has been serving the food, beverage, pharmaceutical, nutritional supplement, cosmetic, and specialty chemical industries for more than 25 years.

What you gain with ProcessPro is the experience and insight of process manufacturing experts who understand the unique needs of your industry, from lot tracking to formulation management to regulatory compliance.

SAP

SAP does not need much of an introduction. They are the largest vendor of enterprise software applications for supply chain management. SAP has multiple products that are showcased in this book, including SAP ERP and SAP SCM/APO SNP and SAP SCM/APO SPP.

www.sap.com

Author Profile

Shaun Snapp is the Founder and Editor of SCM Focus. SCM Focus is one of the largest independent supply chain software analysis and educational sites on the Internet.

After working at several of the largest consulting companies and at i2 Technologies, he became an independent consultant and later started SCM Focus. He maintains a strong interest in comparative software design, and works both in SAP APO, as well as with a variety of best-of-breed supply chain planning vendors. His ongoing relationships with these vendors keep him on the cutting edge of emerging technology.

Primary Sources of Information and Writing Topics

Shaun writes about topics with which he has first-hand experience. These topics range from recovering problematic implementations, to system configuration, to socializing complex software and supply chain concepts in the areas of demand planning, supply planning and production planning.

More broadly, he writes on topics supportive of these applications, which include master data parameter management, integration, analytics, simulation and bill of material management systems. He covers management aspects of enterprise software ranging from software policy to handling consulting partners on SAP projects.

Shaun writes from an implementer's perspective and as a result he focuses on how software is actually used in practice rather than its hypothetical or "pure release note capabilities." Unlike many authors in enterprise software who keep their distance from discussing the realities of software implementation, he writes both on the problems as well as the successes of his software use. This gives him a distinctive voice in the field.

Secondary Sources of Information

In addition to project experience, Shaun's interest in academic literature is a secondary source of information for his books and articles. Intrigued with the historical perspective of supply chain software, much of his writing is influenced by his readings and research into how different categories of supply chain software developed, evolved, and finally became broadly used over time.

Covering the Latest Software Developments

Shaun is focused on supply chain software selections and implementation improvement through writing and consulting, bringing companies some of the newest technologies and methods. Some of the software developments that Shaun showcases at SCM Focus and in books at SCM Focus Press have yet to reach widespread adoption.

Education

Shaun has an undergraduate degree in business from the University of Hawaii, a Masters of Science in Maritime Management from the Maine Maritime Academy and a Masters of Science in Business Logistics from Penn State University. He has taught both logistics and SAP software.

Software Certifications

Shaun has been trained and/or certified in products from i2 Technologies, Servigistics, ToolsGroup and SAP (SD, DP, SNP, SPP, EWM).

Contact

Shaun can be contacted at: shaunsnapp@scmfocus.com

Abbreviations

BOM—Bill of Materials

DRP—Distribution Requirements Planning

ERP—Enterprise Resource Planning

MES—Manufacturing Execution Scheduling

MRP—Material Requirements Planning

NRV—Net Realizable Value

RPM—Rotations Per Minute

www.ingramcontent.com/pod-product-compliance
Lightning Source LLC
LaVergne TN
LVHW080039210926
848481LV00011B/3235